J16.—

G·HOUSER.

THE BOATSWAIN'S MANUAL

THE BOATSWAIN'S MANUAL

BY

WILLIAM A. McLEOD

REVISED BY

CAPTAIN A. G. W. MILLER

GLASGOW
BROWN, SON & FERGUSON, LTD.
4–10 DARNLEY STREET

First Edition	-	-	- 1944
Second Edition		-	- 1957
Third Edition	-	-	- 1968
Revised	-	-	- 1972
Revised	-	-	- 1976
Fourth Edition		-	- 1977
Revised	-	-	- 1984

ISBN 0 85174 475 3   Fourth Edition Revised

© 1984 Brown, Son & Ferguson, Ltd., Glasgow, G41 2SD

Printed and Made in Great Britain

PREFACE

Most seamanship books available to young seamen are of the advanced text-book type, intended for candidates studying for Certificates of Competency as Mate or Master and dealing with the technical subjects connected with their Examinations.

Such books are of little use to those who wish to learn the rudiments of seamanship work. For this reason, these pages have been compiled as a book of elementary knowledge for the beginner and also as a reference-book for older deck ratings, especially those who seek information relating to everyday seamanship problems, or who wish to prepare for the qualifying examination for the Certificate of Qualification as Efficient Deck Hand (E.D.H.) or for the Certificate of Competency as A.B., issued by the Department of Trade.

The Regulations relating to the issue of these certificates are included so that the seaman may appreciate the practical value of the information included in the book. The Chapters are arranged to deal with general ship equipment, manual seamanship, mechanical and cargo gear, tanker knowledge, boatwork, cleaning gear, painting, miscellaneous hints and navigational equipment in roughly alphabetical order and should be useful to executive officers and others for reference purposes. Metrication is still slow to come into everyday use and Imperial units are retained except in areas such as paint and rope where the manufacturers have given a lead.

ACKNOWLEDGEMENTS

The changes in ships and seamanship continue and some men will complete their careers at sea in ships where very little of the old style seamanship is required. "M" Notices and British Standards leave little scope for the personal touch but I hope that there is still an interest in the job which goes beyond mere regulations.

I am grateful to Messrs. D. Thomas for reading the work and to D. Rogers for his illustrations.

A.G.W.M.
Gravesend, 1983.

CONTENTS

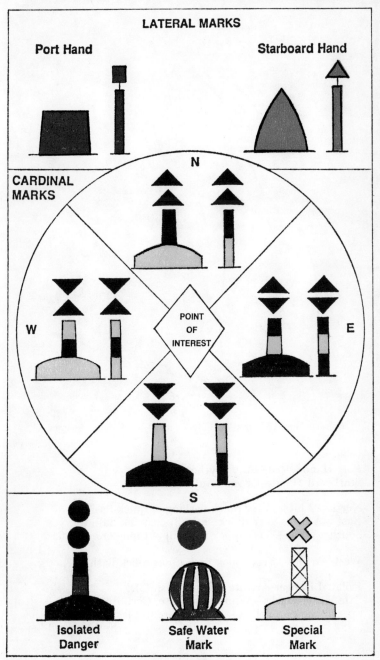

System A Buoys.

Conventional Direction of Buoyage for U.K. Waters—System A.

Lights.

    Port Hand—Red—Any rhythm.
    Starboard Hand—Green—Any rhythm.

    North—White—Very quick flashing or quick flashing.
    East—White—Very quick flash (3) or quick flash (3).
    South—White—Very quick flash (6) +1 long or quick flash (6)
                                        +1 long.
    West—White—Very quick flash (9) or quick flash (9).

    Isolated Danger—White—Group flash (2).
    Safe Water-Marks—White—Isophase, Occulting or long flash.
    Special Marks—Yellow—Flash or Group Flash (3) but not the
                                     rhythms used before.

THE BOATSWAIN'S MANUAL

CHAPTER I.

INTRODUCTION

Types of Vessels.   Certificates :- A.B., E.D.H. and Lifeboatman

Foy Boats.—An ordinary type of pulling boat used for attending ships at anchor or when docking in the North East coast ports of England.

Dorys.—Various types are to be found in different parts of the world, but the best known are those used by the fishermen of the Newfoundland Grand Banks.   Although they are flat bottomed they are excellent sea-boats.

Pram.—A peculiar type of boat without the usual style of stem. Having no forefoot, it tows well behind small sailing yachts, and that is really the purpose for which it is built.

Dinghy.—Small light-weight boats for rowing or sculling and generally used by yachtsmen as tenders.   (*See* Yachts.)

Jolly-boat.—In addition to the lifeboats required by law, most tramp steamers carry a small substantially built boat which is used for all general purposes.   Buoyancy tanks are not usually fitted.

Whale Boat.—A type evolved by the whalermen of the old days. At present they are only found on naval vessels and yachts, and are good boats for both sailing and rowing.

Surf Boat.—Vessels trading on the West Coast of Africa always carried these boats, so that cargo could be "run" through the surf and landed on open beaches.   All were high at the ends and low in the "waist".

Ship's Lifeboats.—These are specially built and equipped under Department of Trade supervision for the purpose of saving life at sea.

Cutter.—Small, sharp built, but broad-beamed vessel carrying a single mast with fore and aft mainsail, gaff topsail, stay foresail and jib.

Yawl.—Is much the same as a cutter, but carries a mizzen mast stepped well aft.

Ketch.—The only difference between a ketch and a yawl is that the mizzen mast in the ketch is stepped forward of the rudder head.

Sloop.—A simple fore and aft rig with a mainsail and one staysail.

Bermuda Rig.—A popular rig with yachtsmen.   The large peaked mainsail does away with a gaff topsail.

1

SILHOUETTES

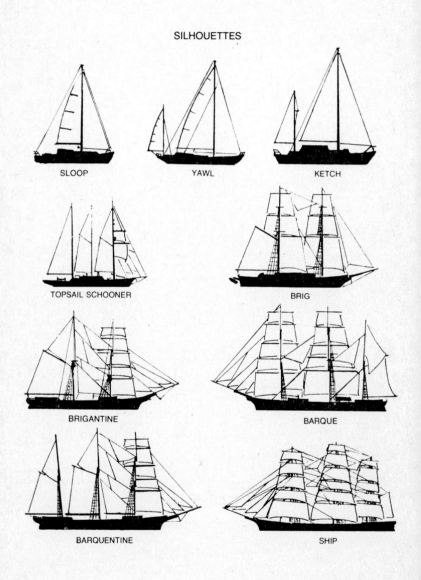

SLOOP

YAWL

KETCH

TOPSAIL SCHOONER

BRIG

BRIGANTINE

BARQUE

BARQUENTINE

SHIP

Fore-and-aft-Schooner.—While some vessels of this type have two or three masts, others have as many as six.

Topsail Schooner.—A two or three-masted vessel carrying topsails, the one on the foremast being square-rigged.

Staysail Schooner.—In this type all fore-and-aft sails are used without gaffs.

Brig.—Two-masted square-rigged vessel carrying a boom mainsail.

Brigantine.—Two-masted vessel, square-rigged on the foremast and fore-and-aft rigged on the mainmast.

Barque.—Three or four masts. Square-rigged on all except the after one.

Barquentine.—Three or four masts, but square-rigged on the foremast only.

Full Rigged Ship.—Three-masted vessel, square-rigged on each mast.

Thames Sailing Barge.—A type of vessel peculiar to the River Thames. They are sprit-sail rigged.

Dumb Barge.—An ordinary barge without propelling power, except long sweeps or oars.

Cruisers.—Apart from the remaining Aircraft Carriers, the Cruiser is now the largest surface fighting unit and, in addition to having a high turn of speed, is heavily armed and armour-plated.

Auxiliary Cruisers.—Fast armed merchant vessels commissioned by the Admiralty in time of war.

Aircraft Carriers.—Large ungainly craft whose chief characteristic is a large wide deck for aircraft to land upon, and for the convenience of which the funnel is placed at the side of the vessel.

Side Trawler.—There is practically no variety in the design of side trawlers and very little to distinguish them from drifters. Both have their engines and wheelhouses aft to allow working space on the foredeck. To work the trawl over the side, there are two heavy inverted U fairleads called gallows, and these are the best guide to the type of fishing carried out.

Drifter.—The drifter has its nets over the bow and needs a big capstan forward to pull it up to the nets, which are to windward. To keep it head to wind, the foremast is lowered and a mizzen sail is set.

Stern Trawlers.—Working on the foredeck of a side trawler is very dangerous, and modern trawlers are built with the wheelhouse forward and a covered work-space aft. The trawl is brought up into the stern on a sloping ramp.

Ice Breakers.—These vessels vary in size and power, but they are all designed to ride up on to the ice and crush a way through for conventional ships to follow.

SILHOUETTES

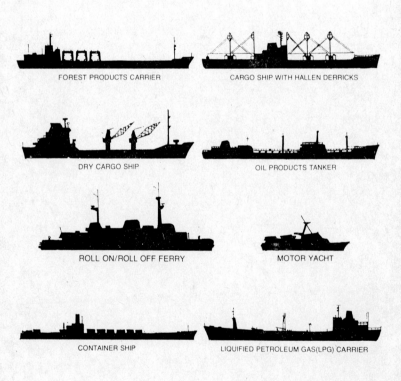

VERY LARGE CRUDE OIL CARRIER

BULK DRY CARGO CARRIER

FOREST PRODUCTS CARRIER

CARGO SHIP WITH HALLEN DERRICKS

DRY CARGO SHIP

OIL PRODUCTS TANKER

ROLL ON/ROLL OFF FERRY

MOTOR YACHT

CONTAINER SHIP

LIQUIFIED PETROLEUM GAS(LPG) CARRIER

Yachts.—Most pleasure yachts are powered by diesel engines, the steam yacht with its big crew is a thing of the past. Sailing yachts can be seen in their thousands on some parts of the coast, and the popularity of the sport for both cruising and racing is apparent to any observer on coastal or inland waters.

Whale Chasers.—Small, fast steamers which work in conjunction with Whale Factory Ships, and which have an outline very similar to that of a trawler. These vessels are easily distinguished by the harpoon-gun right forward, and a flying bridge leading from the gun to the wheelhouse.

Tug Boats.—Small craft, with high-powered engines, which perform all towing operations. All are a slight variation of one type and are easily identified. The main feature is that the after-end is kept clear of all obstructions likely to interfere with the tow-rope.

Cable Ship.—The general outline and colouring are much the same as a steam yacht. The clipper type bow is adapted for handling deep-sea cable, and is fitted with large wheels or guides, which are fitted into the stem head and poop for that purpose.

Hoppers.—Special type vessels employed in the carriage of mud and sand from dredgers to the open sea, where it is discharged by the simple process of opening doors in the bottom. This is done by a system of chains running along the centre-line girder, which forms a very noticeable feature in these vessels.

Dredgers.—Bucket dredgers cannot disguise their occupation as the buckets capsize to empty several tons of spoil before returning to the bottom in an endless chain. They are always attended by hoppers. Suction dredgers are not always obvious, as they may be converted from other trades, such as colliers. A large diameter pipe hinged forward trails diagonally back under the ship and pumps suck up spoil and water. The heavy spoil sinks into the hold and the water overflows over the hatch coamings. Suction dredgers do not need hoppers, and some are engaged to bring sand and gravel ashore for the building industry.

Supply Boat.—A special ship built to service oil rigs in exposed positions off shore. The cargo is carried on deck aft and to keep the after deck as clear as possible the wheelhouse and funnels are forward of amidships. Additional duties may include towing rigs and barges and picking up anchors.

Cross-Channel Vessels.—These are fast vessels originally connected with the train services for the carriage of passengers and mails across short sea routes. They include vessels specially constructed as "Train Ferries" whereby sections of rolling stock can be directly shunted aboard for the journey to the Continent and also vessels constructed as drive-on "Car Ferries" with entrance to car decks through stern doors.

SILHOUETTES

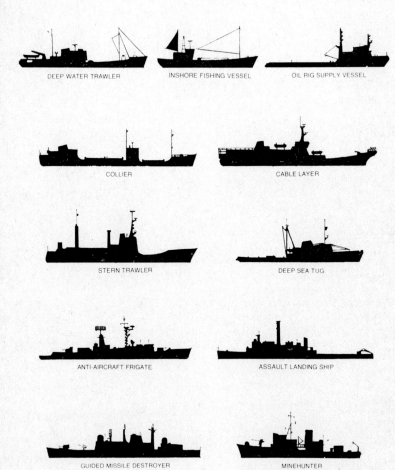

DEEP WATER TRAWLER INSHORE FISHING VESSEL OIL RIG SUPPLY VESSEL

COLLIER CABLE LAYER

STERN TRAWLER DEEP SEA TUG

ANTI-AIRCRAFT FRIGATE ASSAULT LANDING SHIP

GUIDED MISSILE DESTROYER MINEHUNTER

Container Ship.—Vessels specially constructed for the carriage of cargo in large containers stowed in cellular arrangements in the holds and two or three deep as deck cargo. This type of vessel is easily recognisable.

Stern Wheelers.—Light draught steamers operating mostly in shallow tropical rivers and propelled by a paddle wheel fitted across the stern.

Whale Factory Ships.—Steamers fitted with a special slipway over the stern, so that dead whales brought alongside by the "whale chasers" may be hauled on board and reduced to oil in special tanks. The funnels are always placed at the sides of the vessel to provide a clear run for the slipway.

Timber Carrier.—Vessels employed regularly in the timber trades can be distinguished by their high derricks. To allow for a high deck cargo, the masts and winches may be on the "islands".

Collier.—The predominant feature is always the absence of derricks in order that cranes and other mechanical means may be used for loading and discharging the coal. To minimise the trimming of coal out to the "wings" of the hold the hatches are usually made as wide as possible.

"Goal Poster".—Some of the larger cargo vessels of the flush deck variety have the derricks hung on a series of samson posts. These are placed abreast and connected at the top. They are similar in outline to goal posts—hence the name.

"Three Island" Ship.—So called on account of the three raised sections—forecastle head, centre castle and poop. It is perhaps the commonest type of all cargo vessels and typical of tramp vessels of all nations. Another name very often heard is "well deck ship".

Ro-Ro ("Roll On Roll Off").—No cargo gear is needed on these ships as the cargo arrives at the dock on a lorry or trailer, which rolls on and off the ship to its destination.

O.B.O. ("Oil Bulk Ore Carrier".)—So sub-divided that oil can be carried in the large compartments; whilst on the return voyage, ore which is very heavy, may be carried in the small compartments. After proper cleaning the entire capacity could be used for a lighter cargo, such as grain.

Mini-Bulker.—Now that few colliers are engaged in the coal trade in European waters, a new type of coaster is to be seen. Its long, wide opening hatches can accept any cargo which can be loaded by chute or conveyor belt and discharged with shore-based grabs or suction.

Oil Tankers.—At one time the flat silhouette of the tanker, with its engines aft was sufficient to identify it. Nowadays bulk carriers and O.B.O.'s have a similar appearance and only when the tank lids and pipework can be seen can there be certainty.

Coasters.—These vessels are no longer a recognisable group. The

1000-ton cargo ship can be seen much further afield than the coast of the United Kingdom and special ships of all types run feeder services which at one time would have been the work of coasters.

Flush Deck Ship.—A vessel without any wells so that the weather deck forms a continuous line from forward to aft but it may have a raised forecastle. A Shelter Deck vessel was similar in appearance but the uppermost deck was of lighter construction.

Clipper Bow.—A term which indicates the curved bow always seen in sailing ships. Once popular in steamship design it is now rarely seen, except, perhaps, in the case of yachts.

Raked Stem.—The ordinary straight stem of a steamer set at an angle and very popular in all classes of vessels. Two principal advantages are derived from this type.

In the event of a collision, most of the damage to the other vessel would be above the water-line. The reserve buoyancy arising from the angled stem also assists a vessel to ride a sea with greater ease and comfort.

Ice Breaker Bow.—Vessels regularly employed in the Baltic and some other trades are generally constructed with a type of bow which enables them to run on the edge of the ice and break it with the downward pressure of the weight of the ship.

Counter Stern.—Although the overhanging counter is of very old design, it is said to be the best type for a vessel in bad weather.

Cruiser Stern.—A rounded stern first seen in naval cruisers and then in merchant ships of the 1930's. It is largely superseded by a square, flat stern which is easier to build.

Motor Torpedo Boats.—Speedboats armed with machine guns and torpedo tubes, capable of doing fifty knots.

Heavy Lift Ship.—Vessels designed for the carriage of rolling stock and heavy machinery. Very substantial masts, derricks and lifting block are the prominent features of this type.

ISSUE OF CERTIFICATES OF COMPETENCY
AS A.B. AND EFFICIENT DECK HAND.

On and after 1st May, 1953, a seaman engaged as an A.B. in a British ship must hold a Certificate of Competency as A.B., and the fact that such a Certificate has been issued is duly recorded in his continuous Discharge Book (Dis. A). The latest Notice No. M.780 of the Department of Trade, takes into account service at sea reckonable as qualifying service which will now include in addition to service as a deck rating, service as a general purpose rating in ships having general purpose or integrated crews. Qualifying service may be reduced in the case of persons undergoing courses of pre-sea training at prescribed training schools and new provisions are made for reductions to apply in the case of persons who have served at sea as an engine-room rating and undergone training in duties of a deck rating.

Ex-engine-room ratings who have attended an approved engine-room/deck conversion course for which a maximum of four weeks' remission may be claimed, may take the qualifying examination at the end of the course which also applies to courses at certain other approved establishments. However, no certificate will be issued until all other conditions have been satisfied when application should be made for the issue of an E.D.H. certificate. Service as an engine-room rating may count towards qualifying service at the rate of one half of that service up to a maximum of six months, and counts as equivalent to deck rating service.

Conditions for Issue of Certificate.

Certificates are issued on application on Form Exn. 50B to seamen proving service on or before 1st May, 1952, as an A.B. or equivalent or superior deck rating including certificated Efficient Deck Hands, Quartermasters, Bosuns, Bosun's Mates, Deck Store Keepers, fourth year Apprentices and Uncertificated Deck Officers. Apart from the above, the conditions for an award of a Certificate of Competency as A.B., are that a seaman must make application on Form Exn. 50A and prove to the Superintendent of a Mercantile Marine Office that he has:—

(i) *passed the qualifying examination;*

(ii) *attained the age of 18 years;*

(iii) *performed the appropriate sea service;*

(iv) *is in possession of a Certificate of Efficiency as a Lifeboatman, though this examination may be taken at the same time as the qualifying examination.*

(v) *is in possession of a Steering Certificate.*

Qualifying Examination.

This serves as the examination for the award of a Certificate of Qualification as Efficient Deck Hand and is open to all seamen of 18 years or over providing they have performed:—

(a) 12 months at sea as a deck rating; or

(b) 18 months as a general purpose rating; or

(c) between 12 and 18 months for mixed service, such periods to be counted proportionately.

At least 25 per cent of these periods must have been performed in ships (not fishing boats) of not less than 100 gross tons and the remainder of the service may be performed in any type of vessel of 15 gross tons or more but will only count at half rate if under 100 gross tons or in sailing ships of less than 40 gross tons. Service in sea-going vessels of the Royal Navy is allowed subject to the above condition.

Candidates must also produce a Steering Certificate.

Steering Certificate.

Candidates for the qualifying examination will be required to produce a Steering Certificate to show that, apart from periods of instruction, he has taken turns at the wheel for periods totalling at least 10 hours in steering ships (not fishing boats) of 100 gross tons or more. The Master of each ship in which turns at the wheel are taken should certify the time spent until the minimum period is reached. The Steering Certificate (Form Exn. 50G) is issued by the Master to candidates requiring them.

Application to take the Qualifying Examination.

Applications are made at a Mercantile Marine Office on Form Exn. 50C, but in the case of a seaman who is eligible in all other respects for the immediate issue of a Certificate as A.B., the application should be made on the combined Form Exn. 50D. The completed form will be lodged with the Superintendent who will arrange for the examination to be taken as soon as possible at one of the following ports (as named by the seaman in his application):—
Aberdeen, Belfast, Blyth, Bristol, Cardiff, Falmouth, Glasgow, Great Yarmouth, Grimsby, Hull, Leith, Liverpool, London, Middlesbrough, Newcastle, Southampton, Sunderland and Swansea.

The examination (oral only) will be based on the following syllabuses:—

Nautical Knowledge.

1. The meaning of common nautical terms.
2. The names and functions of various parts of a ship, for example decks, compartments, ballast tanks, bilges, air pipes, strum boxes.
3. Knowledge of the compass card 0° to 360°. Ability to report the approximate bearing of an object in degrees or points on the bow.
4. Reading, streaming and handing a patent log.
5. Markings on a hand lead line, taking a cast of the hand lead and correctly reporting the sounding obtained.
6. Marking of the anchor cable.
7. Understanding helm orders.
8. The use of lifesaving and fire-fighting appliances.

Practical Work.

(To be tested as far as possible by practical demonstrations.)

9. Knots, hitches and bends in common use:

Reef knot	Bowline and bowline on the bight
Timber hitch	Sheet bend, double and single
Clovehitch	Sheepshank
Rolling hitch	Round turn and two half hitches
Figure of eight	Marlinspike hitch
Wall and crown	

To whip a rope's end using plain or palm and needle whipping. To put a seizing on rope and wire. To put a stopper on a rope or wire hawser, and derrick lift.

10. Splicing plaited and multi-strand manila and synthetic fibre rope, eye splice, short splice and back splice. Splicing wire rope, eye splicing using a locking tuck. Care in use of rope and wire.
11. Slinging a stage, rigging a bosun's chair and pilot ladder.

12. Rigging a derrick. Driving a winch, general precautions to be taken before and during the operation of a winch whether used for working cargo or for warping.

13. The use and operation of a windlass in anchor work and in warping. Safe handling of moorings with particular reference to synthetic-fibre ropes and self-tensioning winches. Precautions to be taken in the stowage of chain cable and securing the anchors for sea.

14. A knowledge of the gear used in cargo work and an understanding of its uses. General maintenance with particular reference to wires, blocks and shackles.

15. The safe handling of hatch covers including mechanical hatch covers, battening down and securing hatches and tank lids.

16. If no lifeboatman's certificate is held, a candidate will be required to satisfy the examiner that:

    (a) he understands the general principles of boat management and can carry out orders relating to lifeboat launching and operation and the handling of a boat under sail;

    (b) he is familiar with a lifeboat and its equipment and the starting and running of the engines of a power boat;

    (c) he is familiar with the various methods of launching liferafts and precautions to be taken before and during launching, methods of boarding and survival procedure.

Issue of Certificate as Efficient Deck Hand.

Successful candidates at the qualifying examination who do not qualify by length of service for the A.B. Certificate will be issued with an Efficient Deck Hand's Certificate on production of Form Exn. 50E and their Discharge Book at any Mercantile Marine Office and payment of a fee.

CERTIFICATE OF EFFICIENCY AS LIFEBOATMAN.

Lifeboatman's Certificates are issued by, or under the authority of the Department of Trade, and any member of the crew may act as a lifeboatman provided he holds such a certificate.

Applicants must be not less than 18 years of age and shall be free from any physical defect or other disability which might render them unfit for the duties of a lifeboatman.

Pursers, radio operators, engineer officers, surgeons and any other officer ranks must prove not less than six months' sea service before they will be examined and not less than one year's sea service is required in the case of deck ratings, firemen, stewards and other candidates.

Merchant Shipping Notice No. M.579 gives the following detailed syllabus for the Examination:—

(a) Each candidate submitting himself for examination for the issue of a Certificate of Efficiency as Lifeboatman will be required:—

    (i) to identify the permanent markings on a lifeboat and liferaft with regard to the number of persons to be carried.

    (ii) to know the statutory equipment required in a lifeboat and liferaft and its correct use.

(iii) to know the location of instructions provided with particular items of equipment, for example transmitter, first aid outfit, pyrotechnics. (A detailed knowledge of these instructions will not be required.)

(iv) to know the minimum of food and water required for each person the boat or liferaft is certified to carry.

(b) Candidates will be required to demonstrate a knowledge of the following:—

(i) Boat Launching and Handling—Theory.

The emergency signal, its meaning and the action to be taken on hearing the signal.

The difference between the emergency signal and the abandon ship signal.

The procedure to be followed in the preparation, swinging out, embarking, lowering and launching of a lifeboat and the dangers attending these operations.

The procedure from the launching of a lifeboat to leaving the ship and riding to a sea anchor.

The management of a boat under sail.

(ii) Boat Launching and Handling—Practical.

Each candidate will be required to demonstrate his ability to row, steer and act as coxswain during launching operations and afloat, and to show that he understands all the orders commonly used during boat handling. He will also be required to assist in setting the sail, floating the sea anchor and the management of a boat under sail.

(iii) Lifeboat Engine Starting Procedure.

Fuel on.

Gear lever neutral.

Throttle set.

Operate decompression device.

Turn engine over with starting handle and when turning well, change to full compression.

When engine is running, check that cooling water is circulating and that oil pressure is satisfactory (if gauge is fitted).

(iv) Liferafts.

Methods of launching and the precautions to be taken before, during and after launching.

Boarding from the ship or from the water.

Righting an inverted liferaft.

(v) Survival Procedure, Lifeboats and Liferafts.

Rigging of protective covers.

Comfort of passengers and crew.

Issue of food and water.

Maintenance of discipline.

CHAPTER II.

GENERAL SHIP EQUIPMENT

Ship Fittings and Parts. Anchors and Cables. Time and Watch Bells. Steering Gear. Deck Machinery.

Fig. 1.—*See* Folding Plate.

Anchor Cables.—An anchor cable is made up from a number of lengths of chain cable shackled together. Each length is 15 fathoms long, but the entire length and size of the cable varies with the size of the ship, and is governed by Lloyd's Rules and Regulations. A small vessel might only have about 100 fathoms, whereas, a large vessel may carry 300. Similarly, the metal in the links may be anything from one to three inches in diameter or even more.

Cable stowed in the bottom of the locker may become brittle through lying idle. Occasionally, when the vessel is in dry-dock, two or three lengths at the end of the cable are exchanged with a similar number near the anchor to help keep all the cable in good condition.

Shackles.—The joining shackles of a cable are placed so that the bow, or round end, runs first, and those which join the cable to the anchor are larger than the others, but in both types the pin does not project beyond the widest part of the shackle. To secure the pin, a tapered hardwood plug is hammered into a hole through the lugs and pin as well. Sometimes a metal plug is used, in which case a lead plug is hammered in on top of it to hold it in place. Fig. 00, page 103.

Patent Shackle.—A three part jigsaw which when fitted together is the same shape and size as a link in the cable.

Shackle Marks.—By marking links close to a shackle, the amount of cable in the water is easily known. Each link, except the end ones, has a stud across the widest part to prevent the chain "kinking" but on the end one it is left out, so that the shackle can be passed through more easily. When marking the cable, only the studded links are counted, therefore, at the first shackle, the first studded link on each side has a piece of wire twisted round the stud, and at the second shackle the second studded link is marked, and so on. It is a good plan to paint the links white, so that they are easier to see when flying out. Each shackle, of course, represents 15 fathoms. Therefore, if six shackles have run out of the hawse pipe, there will be 90 fathoms of cable in the water.

Securing the End.—The inboard end of the cable is made fast to the bottom of the chain locker by means of a heavy eyebolt and a shackle or a Senhouse slip. In some ships the end is made fast to the top of the locker, and is secured to a heavy lug by means of a strong pin, in which case it is much easier to "slip".

Devil's Claw (or Dogs).—A claw arrangement on the fore part of the windlass which fits over the link of the cable. It is connected

FIG. 2.—Devil's Claw.

to a lug on the deck with a bottle screw for heaving it tight. It acts as a lashing on the cable by taking the weight of the anchor off the windlass while a vessel is at sea. A small length of chain is sometimes used instead of the actual claw. Fig. 2.

Anchors.—Lloyd's Rules and Regulations govern the weight of anchors used on all vessels and provide that the maker's name or initials, progressive number and weight must be marked on each one.

Bower Anchor.—All vessels are required to carry two bower anchors and a spare bower. Sometimes the spare bower is kept on the forecastle head, where it can be picked up and lowered into place with a shore crane. When carried on the foredeck it is handled with a cargo derrick.

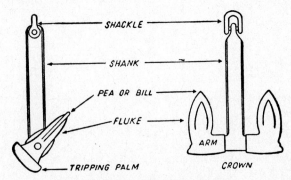

FIG. 3.—Patent or Stockless Anchor.

Patent Anchors.—Practically all vessels use the stockless or patent bower anchors. They consist of a shank attached to a movable head on which are the flukes. The head must weigh not less than three-fifths of the total weight of the anchor. When hove up the shank is right up inside the hawse pipe, the flukes fitting close to the bow plates. Fig. 3.

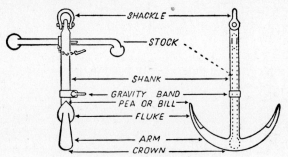

FIG. 4.—Stream Anchor (Admiralty Pattern).

Stream Anchors.—When a vessel is anchored in narrow waters, a stream anchor is sometimes used to prevent her stern swinging about. The handiest place for it is right aft, so it was always kept in that place. Some vessels are fitted with a stockless or patent stream anchor, but the majority use the type fitted with a stock, which must be one-fourth the specified weight of the anchor. A stream anchor is about one-third the weight of a bower anchor. Fig. 4.

Anchor Buoy.—Anchors often have a small buoy attached. It serves the double purpose of showing the master the position of the anchor, and is a valuable guide in the recovery of "lost" anchors.

Grapnel.—This might be described as a small anchor. It is simply a shank with a ring at one end and about six curved flukes at the other. Small ones with a length of line bent on are useful as a hookrope when picking up an anchor buoy. Fig. 5.

Mushrooms.—A type of anchor used as a rule for permanent moorings, particularly for buoys. They are mushroom shaped, and the cable is shackled to the ring in the stem. Fig. 6.

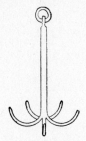

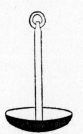

FIG. 5.—Grapnel.   FIG. 6.—Mushroom.

Anchor Light Halliards.—Vessels at anchor exhibit two anchor lights (one, if under 50 metres in length). One is hoisted with halliards rove through a block on the fore topmast stay, while the other is hoisted on the ensign staff aft.

Baggage Room.—An enclosed space given over to passengers baggage which may be wanted occasionally during a voyage.

Ballast Tanks.—As an aid to stability and trim, ballast tanks are a necessity in every ship. Their most important function is to increase the stability of vessels which have no cargo on board. With all the tanks full, a vessel will sit deeper in the water and she will be more seaworthy, the propeller will be deeper in the water and more effective.

Practically every vessel is built with a double bottom divided into sections, and water ballast is carried in the space thus provided. In addition to this, there are usually fore peak and after peak tanks, both of which are situated in the places indicated by their names.

Belaying Pin.—An iron pin found on the sheerpole for making ropes fast in the same way as a cleat.

Bells.—All vessels have to be provided with an efficient bell, which must be hung in some place clear of all obstructions, must be not less than 300 mm. diameter at the mouth.

See Annex III. Regulations for Prevention of Collision at Sea.

These regulation bells are always carried on the forecastle head, but a smaller one will be found on the bridge, and in the crow's nest also.

Time Bells.—Throughout the full twenty-four hours, the bridge bell was struck every half hour to indicate the time to the watch. The look-out man on the forecastle or in the nest answered it and particularly in coal burners, the engine room might repeat the sound as well. It is well to recall that in steamships there is almost complete silence at night. In modern ships particularly motor ships with a great deal of noise and vibration and small crews, the bells may not be struck at all. Nonetheless when calling the next watch it is usual to inform him that it is 'one bell' and if a job is to be started at 0800 the crew will have a 'seven bell breakfast'

TIME AND WATCHES BY BELLS.

Watches	Time	Bells	
	2000	8	Change of Watch
	2030	1	
	2100	2	
First watch	2130	3	
2000 - 2400	2200	4	
	2230	5	
	2300	6	
	2330	7	
	2345	1	Call watch below

Watches	Time	Bells	
	Midnight	8	Change of Watch
	0030	1	
	0100	2	
Middle watch	0130	3	
2400–0400	0200	4	
	0230	5	
	0300	6	
	0330	7	
	0345	1	Call watch below
	0400	8	Change of Watch
	0430	1	
	0500	2	
Morning watch	0530	3	
0400–0800	0060	4	
	0630	5	
	0700	6	
	0730	7	
	0745	1	Call watch below
	0800	8	Change of Watch
	0830	1	
	0900	2	
Forenoon watch	0930	3	
0800–1200	1000	4	
	1030	5	
	1100	6	
	1130	7	
	1145	1	Call watch below
	Noon	8	Change of Watch
	1230	1	
	1300	2	
	1330	3	
Afternoon watch	1400	4	
1200–1600	1430	5	
	1500	6	
	1530	7	
	1545	1	Call watch below
First dog-watch	1600	8	Change of Watch
1600–1800	1630	1	
	1700	2	
	1730	3	
	1800	4	Change of Watch
Second dog-watch	1830	1	
1800–2000	1900	2	
	1930	3	
	1945	1	Call watch below
	2000	8	Change of Watch

Reporting Lights.—When the look-out man sights a light at night he reports it to the bridge by striking the bell, once if the light is on the starboard side, twice if it is on the port side and three times if it is right ahead. In some ships the reporting is done by telephone.

Relief Bell.—To call the attention of the officer of the watch to the fact that he has not been relieved, or that he wishes to be relieved, the look-out man would strike four bells.

Anchor Bell.—As the anchor is being hove in, the mate keeps the master and pilot informed as to how much cable is still out by striking the bell once for every shackle (15 fathoms) out. For instance, if he saw the fifth shackle coming up the hawse pipe, he would strike five bells and the master would know that seventy-five fathoms of cable was still out. When the anchor is aweigh the bell is rung rapidly for a couple of seconds or so.

Fog Bell—When vessels are at anchor during fog the bell or bell and gong (in vessels 100 m. or more in length) must be rung rapidly for five seconds every minute to indicate the ship's position to other vessels.

Fire Bell.—The bell may be rung rapidly for about five or ten seconds or even more, to sound the alarm for fire stations.

Church Bell.—When a Church Service is to be held on a passenger vessel the bell is struck once every three or four seconds for about three minutes. A good effect is obtained if the small bridge bell is struck during the interval.

Visitors Ashore.—Three times, at quarter-hour intervals, the bell is rung rapidly for a few seconds, to notify visitors that it is time to go ashore.

How to Ring.—To ring a big bell rapidly, move the arm away from and back to the person. Do not attempt to move the arm across the body or the hand will describe circles. When reporting a light, or shackles on a cable, the requisite number of strokes must be made singly or in "ones", with a one second interval in between. When striking the time, the strokes must be in "twos". For instance, to make four bells, strike twice, with an interval of one second in between, pause for two seconds, then strike twice again with the one second interval in between.

Although the majority of seafarers use the term "Strike eight bells", it is not strictly correct. The proper phrase is "Make eight bells".

Bell Lanyard.—Bell lanyards can be made in many ways, according to the fancy and ingenuity of the individual. A plain one may be made with ordinary round sennet, finished off with a knot. Sometimes a small eye is worked on one end, by means of which it is attached to the tongue of the bell.

A neater method is to leave the ends loose, space them evenly around the tongue and finish off with a neat whipping.

Bilge.—The rounded part of a vessel's hull where the vertical sides meet the horizontal bottom.

Bilge Keel.—A large, fairly wide plate fitted at right angles to the bilge. Its purpose is to prevent excessive rolling. They are also known as "Rolling Chocks".

Bilges.—Shallow triangular spaces between the frames of a vessel's hull. They are formed by the junction of the double bottom, the flooring of the hold, and the ship's side at the turn of the bilges. Their purpose is to act as a gutter for water which finds its way into the hold

Bilge Pipes.—Water draining into the bilges is pumped out with bilge pipes fitted throughout the length of the ship. The suction end of the pipe in each hold is fitted with a "rose" or "strum box", which prevents solid matter entering the pipe.

Breakwater.—An erection across the forecastle head to break the force of water sweeping over it during bad weather.

Breast Plate.—A triangular strengthening plate in the bows which takes the shape of the angle formed by the ship's two sides, with the stem head as apex.

Bridge Screens.—Painted canvas screens sewed to the rails which surround the top and lower bridges. Modern ships are fitted with wooden screens and glass panels or windows.

Bulkheads.—Steel or wooden partitions separating holds or other compartments. The collison bulkhead is forward near the stem.

Bucket Handles.—Most fire-buckets are made of wood and have rope handles of 3-inch manila. A Matthew Walker knot is worked on each end to hold the handle in position, and the remaining ends (or strands) are laid up again and whipped for a distance of nearly one inch.

Cabs.—Covered-in shelters at the extreme ends of navigating bridges.

Cargo Battens.—See Portable Side Battens.

Cargo Light Booms.—Booms fitted on each side of the masts. On these, cargo lights were hoisted high above the decks and well outboard, for working cargo at night.

Carpenter's Shop.—Small compartment given over to the use of the carpenter. It might be found in any part of a vessel.

Ceiling.—Wooden flooring in the bottom of a hold.

Chains.—Small platform hinged to the ship's side near the bridge. It is fitted with a canvas apron made fast to stanchions, and against this the leadsman leans while heaving the lead for soundings.

Chart Room.—As the name would imply, all charts are kept in this room, in addition to all other navigating equipment, such as chronometers, sextants, etc. It is always situated on the navigating bridge behind the wheelhouse.

Cleat.—A two-pronged fitting to which ropes are made fast.

Coamings.—The built up sides and ends of a hatch, into which the hatch beams are fitted.

Coffer Dams.—Spaces between two bulkheads forming a void separation between the oil tanks and adjacent compartments.

Companionway.—A means of access from one deck to another. Generally a sheltered stairway.

Condenser.—See Engine-room Discharge.

Counter.—The overhanging part of the stern. Substantial lugs are fitted on each side of it, and to these tackles are hooked when removing the propeller.

Cross Bunker Hatch.—It is situated either forward of, or abaft, the funnel. It is built so that the coal it contains will run down into the side bunkers, as in the case of a saddle back.

Crow's Nest.—The sheltered stand on the foremast in which the look-out stands.

Deep Tanks.—Two (or more) very large and deep tanks. Usually built as part of a ship's hold and situated anywhere towards the middle of a vessel. Used in the ordinary way as part of the hold space when not required for liquid cargoes or for water ballast.

Deck-Lights.—Thick glass let into wooden decks above living quarters, as a means of giving light to dark interiors.

Derrick Crutches.—Special erections which support the ends of derricks when lowered down.

Derrick Outreaches.—An erection to be found on most vessels' masts. Their purpose is to provide a means of supporting derricks, placed on each side of, but well away from, the mast. (See Fig. 10).

Derrick Posts.—See Samson Posts.

Docking Bridge.—A small bridge running across the poop from one side to the other. It enables an officer to direct mooring operations with an unobstructed view of what is going on.

Domestic Tanks.—Tanks containing the fresh water used for cooking, washing and drinking.

Donkey Boiler.—Boiler fitted specially for supplying steam for winches, dynamos, etc., while in port. One of the main boilers is fitted for the same purpose if a special one is not provided.

Dowels.—Round wooden plugs sunk in the deck to plug the holes left by the bolts holding the deck planks in place.

Emergency Batteries.—A secondary power supply for radio transmission in the event of main power failure will always be found in the vicinity of the Radio room. If the same compartment is used for the batteries which supply the emergency lighting around the ship then it will be a small room instead of the old style battery box.

Emergency Light Gear.—Since the introduction of electric lights, gear used for hoisting navigation lights is now kept solely for emergency purposes. It consists of two wire guides, each rove through the lugs of a cage, and set up tight between the deck and "forks" on the lamp bracket. The oil lamp is placed inside the cage, and is hoisted by means of halliards, until it runs on the "forks" which hold it steady in position.

Engines, Reciprocating.—These are either compound, triple or quadruple expansion. That is, either two, three, or four cylinders, through all of which the steam passes before being condensed back to water and returned to the boilers. For instance, in a triple expansion engine, the steam (after passing through the boiler stop valve and engine stop valve) would then enter the high pressure, intermediate and low pressure cylinders in succession, and perform the necessary functions on each piston before passing to the next cylinder. Each successive cylinder is larger, to take advantage of the lessening expansion of the steam. When it finally exhausts into the condenser, it is transformed into water by coming in contact with the colder surface of pipes through which sea water circulates.

Engines, Turbine.—A horizontal cylinder through which the propeller shaft passes, and to which a rotor is attached. Steam striking the rotor with its numerous small blades causes the shaft to revolve at a high speed; but, as it only revolves one way, a second turbine must be fitted to the same shaft for use when going astern. High- and-low-pressure turbines are usually fitted to the same shaft.

Engines' Diesel.—Internal combustion engines which operate by compressing a mixture of oil and air. They can only be started with the aid of compressed air, which moves the pistons until the engine operates on its own. This naturally limits the number of startings when manoeuvring in harbours. On account of this, the D.T.I. Rules require that air tanks shall contain sufficient air for twelve consecutive startings, without being refilled. Many advantages are claimed for this type of engine. For instance, less space is occupied for the installation than for any other type. The oil fuel requires less space, fewer hands are carried, and more cargo can be loaded, and so on.

Engines, Smoke Identification.—Dense black, or even lighter-coloured smoke coming from a ship's funnel will indicate the coal burner, while a light-hued smoke, with a hazy or greasy appearance, is usually emitted by oil-burning vessels. Motor ships are easier to distinguish, as the hazy exhaust shoots out in a rapid series of puffs, in the same manner as a motor-car exhaust.

Engine-Room Discharge.—A noticeable feature of all power vessels particularly when under way, is the large flow of water to be seen coming from an opening on the starboard side amidships, near the load waterline. This comes from the general cooling system which also condenses the exhaust steam into water again for pumping back into the boilers.

Engine-Room Plates.—Steel plates forming the flooring of the engine-room.

Exhaust Pipes.—Steam exhaust pipes are those which run up the after end of the funnel. Motor ship exhausts run inside the funnel.

Eyebolts.—These are usually found at any point where it is necessary to shackle a guy fall.

Eyebrows (or Spurnwater).—Metal channel fitted to the top half of a porthole, so that rain- and sea-water will run clear.

Fairleads.—If a rope habitually leads over any permanent obstruction, a contrivance shaped to minimise damage in the way of chafe, and known as a fairlead, is always fitted in that particular spot. Those fitted at the break of the poop, or forecastle head, usually have a horizontal roller.

Fashion Plates.—The cut-away plates at the corners of the well decks.

Feeder.—Wooden trunkway fitted around a hatchway, between two decks. When a grain cargo settles and leaves a space on top, the space is automatically filled with the weight of grain lying in the feeder.

Fiddley.—The casing over the engine and boilers upon which the funnel stands. It also refers to the gratings and ladders which led from the main deck down to the stokehold plates.

Fish Plate.—Boundary plate on the outboard side of each deck above the main deck. Rails and stanchions on the saloon deck are riveted to it.

Forefoot.—The rounded part where the stem post meets the keel.

Flying Bridge.—The decks of oil-tankers are so cluttered with pipes and other obstructions and so much water is shipped when in heavy seas that a flying bridge is provided well above the decks for the safe passage of the crew from forward to aft and vice versa.

Frames.—Vertical angle bars or girders extending from the floors at the bottom of the vessel to the beams at the top and together forming a transverse member giving shape to the ship. The frames are in effect the ribs of the vessel to which the shell plating is riveted. *See also* "Marking Plates and Frames".

Fresh-Water Tanks.—After all loose rust and sediment has been removed, the inside is given a coat of limewash to purify it and this also has a beneficial effect on the water.

Funnel.—Uptake for smoke from the ship's furnaces. If an inside casing is provided—as it invariably is—heat on the outside casing is considerably reduced. Some are provided with a gallery for working on so that the top may be unshipped for passage up the Manchester Canal.

Funnel Stays.—Usually about 4, 6, or 8 stays give support to the funnel. In hot weather, owing to expansion of the funnel, they often require to be slacked off, and to do this automatically, a spring arrangement is sometimes fitted to the stay itself.

Gallery.—*See* Funnel.

Galley.—Another name for the cook-house.

Galley Funnel.—Uptake for smoke from galley fires. It may be either standing by itself or connected to the main funnel.

Garboard Strake.—The line of plating on each side of the keel. These plates are thicker than the other shell plates.

Goosenecks.—Tank air escapes usually found in the water-ways. Fig. 7. They also act as an overflow, and some are adapted to take a sounding rod. This is also the name for the swivel arrangement which supports the heel of a derrick.

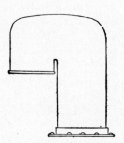

Fig. 7.—Gooseneck Ventilator.

Gunport Doors.—Heavy steel doors in the ship's side used for various purposes, according to their situation. Fruit boats load bananas through them, and passenger vessels use these doors for their gangways, as a means of access to the ship. They will often be found near the galleys and passenger quarters, where they are used solely for ventilation purposes.

Headroom is usually limited when boarding a vessel by means of a gangway landed in a gunport or a foyer door.

As a warning to passengers and others, in an effort to prevent them striking their heads, various remedies are in use. These include notice-boards, red flags, cushions, etc., but one of the best ideas is to use a strip of white cotton canvas or duck, long enough to stretch across the door and having its lower half threaded until a good deep fringe remains. Through fluttering about in the wind, this will attract attention much better than anything else.

B

Hand Gear.—A means of steering a vessel by hand power, when the steam or other type of gear has broken down.

Hatchways.—Openings in the deck giving access to holds bunker spaces and storerooms.

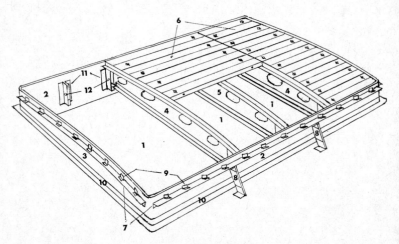

FIG. 8.—Hatchway

1. Hold space.	2. Side coaming.	3. End Coaming.
4. Portable beam (King).	5. Portable beam.	6. Wooden hatches.
7. Stiffeners.	8. Stays.	9. Cleats.
10. Watertight angle iron.	11. Beam carriers.	12. Beam bolt.

Hatch Beams.—These are really heavy steel girders which fit transversely between the coamings of a hatchway.

Hatch Covers.—Heavy wooden covers laid on the hatch beams, to cover the hatchway.

Hatch Cleats.—Iron lugs on the hatch coamings into which the battens are shipped to hold the tarpaulins in place.

Hatch Battens.—Long narrow iron bars which hold the hatch tarpaulins in place and prevent water entering the holds. They are laid in cleats attached to the hatch coamings, and are wedged tightly in place with wooden wedges.

Hawse Pipes.—Steel pipes leading down through the bows in which the anchor cables run. The shank of the anchor fits snugly inside it when hove up.

Holds.—The interior of a vessel is divided up into a number of spaces where cargo is stowed or packed. Each of these spaces is known as a "hold", and is numbered from forward. Fig. 9.

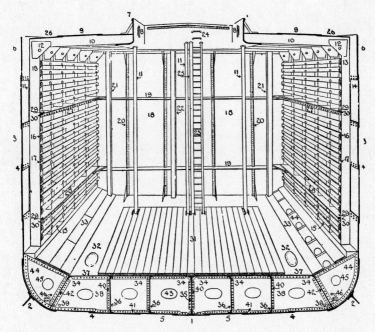

FIG. 9.—Hold.

1. Plate Keel.
2. Bilge Keel.
3. Shell Plating (Raised and Sunken System).
4. Shell or Plate Landings.
5. Garboard Strake.
6. Sheer Strake.
7. Hatch Coamings.
8. Hatch Beam Sockets.
9. Deck Plating.
10. Deck Beams.
11. Hold Pillars.
12. Beam Knees.
13. Frames.
14. Liners.
15. Frame Spaces.
16. Cargo Battens.
17. Cleats.
18. Bulkhead.
19. Transverse Stiffeners.
20. Vertical Stiffeners.
21. Box Protecting Bilge Sounding Pipe.
22. Box Protecting Ballast Tank Sounding Pipe.
23. Hold Ladder.
24. Rungs.
25. Space for Shifting Boards.
26. Deck Stringer Plate.
27. Side Stringer.
28. Bilge Stringer.
29. Intercostal Plate.
30. Angle Bar.
31. Ceiling.
32. Manholes.
33. Limbers.
34. Cellular Double Bottom forming Ballast Tank.
35. Centre Line Longitudinal Girder.
36. Intercostal Side Girder.
37. Inner Bottom Plating.
38. Floor Plate (Transverse Plate).
39. Draining Holes.
40. Vertical Stiffening Angle Bars.
41. Joggled Stiffening Angle Bars.
42. Margin Plate.
43. Lightening and Access Holes.
44. Bilge Space.
45. Tank Side Bracket.
46. Strum Box.

Hounds Band.—An arrangement for securing the rigging to the masthead. In some cases it is a thick iron band riveted to the mast. Eyes are welded on at suitable angles to take the shrouds and stays. Fig. 10 (*see* Folding Plate).

Another system often seen is two angle bars riveted to the mast a few inches apart. These are set at a suitable angle to take the eyes of the rigging, which is held in place with a heavy pin inserted through holes in the angle bars and held in place with a forelock.

Insulators.—On account of electrical disturbance, many vessels, particularly naval vessels and tankers, have their stays fitted with insulators. These are also found on wireless aerials.

Jackstaff.—Flagstaff situated at the stem head.

Jumper Stay.—A wire stay set up between the foremast and the funnel. Signal halliards are always hung from it.

Lamp Brackets.—A bracketed shelf on the foremast, and also on the mainmast, upon which the two mast head lights are bolted.

Lazarette Hatch.—A small storeroom, always at the after end of the ship usually beneath a small hatchway.

Lifebelt Boxes.—Large boxes containing lifejackets placed at strategic points where they can be quickly distributed when required.

Life-Jacket Racks.—Overhead racks fitted in crews' and passengers' quarters.

Bow Locker.—A compartment under the forecastle head near the stem post. As no purpose can be found for what would otherwise be wasted space, it is turned into a locker for all classes of ship's gear. The forepeak hatch is usually in or near the bow-locker.

Chain Lockers.—Two large compartments under the windlass. Into this the port and starboard cables are stowed after coming down through the spurling pipes. If sufficiently large, there is no need to stow the cable, but if not, it must be flaked or stowed, otherwise it will not all fit in.

Flag Locker.—A cabinet standing on legs or screwed to a bulkhead on the navigating bridge. All flags are kept in the cabinet in small pigeon-holes appropriately lettered.

Lamp Locker.—A compartment fitted with the necessary shelves, hooks and tanks for holding all the ship's lamps and illuminating oil. It is always separated from living quarters by an iron bulkhead. Most of the tools required every day are kept in the lamp locker.

Paint Locker.—This may be a proper locker, or just a spare corner set apart for the storage and mixing of paint. Like the lamp locker, it must be separated from living quarters by an iron bulkhead.

Sail Locker.—All awnings, tarpaulins, covers and screens are kept in a sail locker which is always situated in a dry spot, otherwise dampness would be injurious to the canvas.

Sand Locker.—The place where sand is kept is seldom a locker in the sense that the name would imply. Usually, it is just a large box, standing in some position where it is easy to distribute sand to wherever it is required.

"Spud" Locker.—Storage place for potatoes. As a rule it is kept separate from the vegetable locker, but occasionally the two are combined.

Vegetable Locker.—Storage place for vegetables. It is usually situated in an out of the way, cool and airy spot.

Locking Bars.—Iron bars laid across hatchways and padlocked to the coamings. Their purpose is to prevent unauthorised persons entering the hold.

Mail Room.—Well protected storage place where mail is kept during transit in large steamers.

Main Hatch.—The principal hatch where the heaviest weights are usually carried. On the coaming the ship's Official Number and Registered Tonnage are always cut.

Manhole Doors.—Access to tanks and boilers is gained by way of a manhole door. In the case of a boiler, the door is always on the inside, so that the more pressure there is against it, the more tightly it will keep closed.

Both hole and door are elliptical in shape, so that the door may be passed through, and adjusted or withdrawn without any trouble.

The door of a fresh-water tank is usually on the outside, and is held in place with a row of bolts studded into the tank. These engage holes in the cover, and it is eventually screwed tight with nuts.

Packing and white-lead are invariably used to ensure a good watertight joint being made.

Marking Plates and Frames.—By a simple system of lettering and numbering, each plate and frame of a vessel's hull is readily identified.

Beginning with the garboard strake next to the keel, which is A, each strake of plating is marked alphabetically, up to and including the sheer strake which, if it was the fifth, would be marked E.

Individual plates and frames are numbered from aft.

These letters and numbers may often be seen painted on a vessel's side while repairs are being effected in drydock, and they are also used by Masters when reporting damage.

Mast Coat.—A canvas cover which fits around the place where the top mast enters the lower mast. It keeps water out of the lower mast.

Mast Partners.—Plates fitted round the mast hole to strengthen the deck plating.

Mast Step.—The place where the heel of the mast rests inside the ship. At one time it was the custom to step the mast on coins of the realm, but this is now only done in the case of yachts.

Monkey Island.—The deck formed by the top of the wheelhouse and chart house. The standard compass is always found on this deck, and for that reason brass is substituted for iron in the vicinity as much as possible.

Mooring Bitts.—Two wide vertical stumps around which the mooring ropes are made fast.

Mushrooms.—Small ventilators with screw tops, always found on the decks above living quarters. A rubber ring on the inside permits

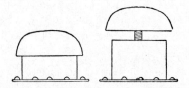

Fig. 11.—Mushroom Ventilator.

them to be screwed tight, to prevent water finding its way down. Fig. 11.

Oertz Rudders.—Increased speed, greater manoeuvring value, as well as lower running costs were claimed for this type of rudder.

Old Man.—A vertical roller on a pedestal, which permits mooring ropes to be led to a winch or windlass, from an awkward angle.

Panama Lead (or Panama Towing Pipe).—The ordinary type of open fairlead is of no use to a vessel in the Panama Canal locks. Mooring ropes have to take such a high angle lead that the Panama Lead was designed for that purpose and is in fact a pipe.

Angle Iron.—Iron bars made to the same shape as the letter L. They are used extensively for joining purposes in the building of riveted ships.

Belay.—To stop heaving and make fast.

Boat Stretcher.—Small wooden spar which fits across the bottom of the boat for an oarsman to brace his feet on. They are to be found in most boats except lifeboats.

Broken Stowage.—Wasted space, or corners into which cases or other cargo will not fit snugly. To counteract this sort of thing, in the case of grain loaded in bags, it is sometimes the custom to "bleed the bags". That is, to slit them, and allow the grain to run out and fill the crevices.

Camber.—The arch of the deck extending from one side of the vessel to the other.

Corrosion.—This only occurs when iron or steel is exposed to the air, and is the principal reason why paint is applied to such surfaces. Oxygen is necessary for the formation of rust, so that is why the air is excluded as much as possible.

"Finished with Engines".—When safely moored, a loud and continued ringing on the engine-room telegraph indicates that the engines will not be required for any further manoeuvring. This is not possible with electric telegraphs which must be moved in regular even steps.

"Glory Hole".—A name often applied to the steward's quarters as a variant of the "steward's peak". The term is said to have arisen through the fact that Nelson died in the cockpit of the *Victory*.

Hogging and Sagging.—A vessel is said to be "hogged" when her ends droop down below the level of the keel, and when the middle section droops she is "sagged".

Isherwood System.—The frames of an ordinary vessel run transversely across the keel, but with the Isherwood system they run fore-and-aft, which gives a vessel great longitudinal strength. Tankers and bulk carriers are usually built on this system or one of its variations.

Light Ship.—Vessel without cargo on board, known as "Flying Light".

List.—It may happen that too much cargo is loaded or discharged on one side, or the continual use of coal or oil from one bunker may cause the vessel to lean over. This would be known as "a list", and may be counteracted with the aid of ballast tanks.

Pitching.—The plunging of a ship's head in a seaway when the ends (bow and stern) go through the sea-saw motion.

Load Water Line.—Water line around a vessel where the surface of the water cuts the ship's side, when she is loaded as deep as the minimum freeboard regulations permit.

Freeboard.—Distance between the waterline and the deck line.

Draught.—The distance from the bottom of the keel to the surface of the water.

Port.—When facing forward the Port side is on the left hand, and Starboard on the right hand.

"Ring Her Away".—A loud and continued ringing on the engine-room telegraph indicates that the voyage proper has begun. In other words, there will be no more stoppings and startings. This is not possible with electric telegraphs which must be moved in regular even steps.

Rolling.—When the sides of a ship perform a see-saw movement she is said to be rolling.

Scending.—A modification of pitching. It occurs when a vessel is carried along on the declivity of a sea, and gives rise to a curious cornerwise motion which conveys the idea of great speed.

Starboard.—When facing forward, Starboard is on the right-hand and Port is on the left-hand.

Superstructure.—A general term indicating all the decks and deck-houses above the main deck.

Trim.—This refers to the way a vessel sits in the water and may be the cause of her steering very badly. The ideal trim is usually "a little by the stern" or, in other words, the ship is deeper in the water aft than she is forward. If she were deeper forward she would be "down by the head".

"Upper Works".—Same as Superstructure.

"Weeping" Joint.—Any joint where rust forms and leaves unsightly red marks running from the affected part is said to be "weeping". This may be stopped by caulking with a chisel.

Yoke Lines.—These are mostly found in yachts' boats. Instead of an ordinary tiller for steering, there is a rudder head which fits across the top of the rudder. Attached on each side are the yoke lines, one of which is held in each hand while steering. The lines are usually of white cotton rope finished off with fancy knots and tassels.

Fore Peak.—The compartment between the stem and collision bulkhead. The lower part is sometimes used as a ballast tank, but the upper section is always the bosun's store, in which most of the rope, paint and cargo gear is stowed.

After Peak.—Waste space adjacent to the stern frame, always utilised as a ballast tank.

Steward's Peak.—Steward's sleeping quarters. In large vessels they are usually berthed near the fore peak or after peak. This compartment is also known as the "Glory Hole".

Platform.—All controls are manipulated from the engine-room platform. It is not a platform in the proper sense of the word, but merely a control centre.

Portable Side Battens or Spar Ceiling.—Substantial wooden spars which prevent cargo coming in contact with the iron frames and plates of a steamer's holds.

Portholes.—Side scuttles is the correct name for the small round openings in the ship's side, more commonly known as portholes. Each is fitted with a "deadlight" hinged over the top. When turned down and screwed tight (by means of the clamps), protection from sea-water is provided if the glass should break during bad weather.

The small rubber band around the edge of both porthole and deadlight should never be painted. This will injure the rubber and cause a leak. If a "port" is screwed up too tight it will crack very easily when a sea hits it. It should be sufficiently tight to stop water coming through and no more, and must be screwed up evenly.

Pump Rooms.—Control rooms for the pumping arrangements in oil tank steamers.

Punka Louvre (Thermotank).—This is a ventilation system whereby fresh air, at any given temperature, can be pumped to any part of the vessel to which it may extend. Air is delivered from a spout working on a universal joint. This permits it to be switched to any part of a compartment.

Quadrant.—Attachment fitted to the top of a rudder post, by means of which the rudder is turned. It takes the shape of a quadrant.

Quarter.—The rounded part of the ship's side aft, on each side of the poop.

Quarter Pipes.—*See* Shoulder Pipes.

Ringbolt.—Heavy iron rings riveted to various parts of the ship and her superstructure, each of which has a particular purpose, such as for guys to be shackled to or for lashing deck cargoes, etc.

Rocket Socket.—A brass socket fitted on each side of a vessel near the bridge. They are for socket signals which are to be fired, and are about 5 inches deep. Sometimes they are screwed to the deck, or may be fitted to the teak rail on the bridge. They are always set at an angle which will throw the signal outboard and clear of the ship. These are now obsolete.

Rolling Chocks.—*See* Bilge Keel.

Rose Box.—The suction end of a bilge pipe terminates in a "rose", and this is protected by a "rose box" or "strum box". It will be found in the bilges in the after ends of the forward holds, and the forward ends of the after holds.

Rudders.—The old-fashioned plate rudder is rapidly going out of use. Modern vessels have streamlined rudders which increase efficiency in steering, and many are fitted with balanced rudders. These have as much area on the fore part of the rudder stock as there is abaft it.

Samson Posts (or Derrick Posts).—Large steel posts, the chief function of which is to support derricks at places where the mast derricks cannot reach. They usually incorporate a ventilator leading to the holds.

Sanitary Tanks.—Large tanks containing water, which are always to be found on the deck above lavatory accommodation, and used for flushing water closets. Modern ships have automatic pumps feeding pressure flushing systems.

Scuppers.—These might be holes cut in the bulwark, where it meets the deck, or they may be pipes which run down from the waterways to near the water's edge, before discharging overboard.

Sheerpole.—An iron bar bolted across the lower ends of all the shrouds. It prevents them from turning and loosening the bottle-screws, and is always fitted with belaying pins.

Sheer Strake.—The line of plating immediately below the bulwarks. Like the garboard strake, these plates are thicker than the other shell plates.

Shell Plating.—The steel plates forming the shell or hull of a vessel. They are riveted to the frames. *See also* Marking Plate and Frames.

Shifting Boards.—When bulk cargoes such as grain are carried, heavy boards are built up in the centre line of the ship's hold, to prevent the cargo shifting to one side when rolling heavily.

Shoulder.—The rounded part of the ship's side forward, near the break of the forecastle head.

Shoulder Pipes.—Oval holes in the bulwarks, at the break of the forecastle head (on each side), through which the backspring is led on to the quay. Another two pipes, situated at the break of the poop (on each side), are known as "quarter pipes", and through these the after backsprings are also led to the quay.

Shrouds.—These are the standing rigging supporting the mast on each side and are usually fitted in pairs. If three are used the forward one would be single and is known as a "swifter".

Side Light Screens.—Screens of wood or metal at one time painted in appropriate colours but now painted matt black. Situated at the bridge wings they hold the red and green electric sidelights. In addition there is either a secondary electric circuit or an oil lamp for emergency use.

Side Pocket Hatches.—Small hatches provided for loading bunker coal in the side bunker spaces immediately underneath. They were always to be found at the sides of a vessel amidships.

Soil Pipes.—Pipes which conduct and discharge soil from the lavatories over the side.

Sounding Pipes.—Narrow pipes leading from the deck to the tanks and bilges. A graduated rod is lowered down inside the pipe to find the amount of water lying therein

Steam Pipe Casings.—Ribbed iron plates, bolted over steam pipes to protect them from injury.

STEERING GEAR.

Hand Gear.—Before the advent of steam, the steering gear of sailing vessels was operated by what is now known as "hand gear". The steering-wheel operated a simple gear which caused the various rudder movements. In fine weather this was easy enough to operate, but during bad weather it proved a heavy job, and often required more than one man to control the helm. Later, with the advent of power, the hand-operated gear became an emergency gear for use only in the event of a breakdown in the main power-operated steering gear. Except in very small vessels it is now obsolete.

Steam Gear.—With the appearance of steam, a new method of steering came into being. All navigation was transferred from the poop to a bridge centrally situated, and, as it would have been difficult to move the heavy lengths of chain and rods to operate the quadrant by hand, steam steering engines came into general use. In this system the turning movement of the wheel on the bridge is communicated by rods and bevel gears to operate the valve controlling a steam-engine situated at deck level in the vicinity of the after part of the main engine-room. The engine is connected by chains and rods along each side of the deck by means of buffer springs to the quadrant.

To help absorb the shocks on the rudder in heavy seas, it was necessary to reeve "relieving tackles" in the form of an endless fall rove through two tackles attached to the quadrant one on each side such that the friction of the rope overhauling itself through the blocks as the quadrant moved, damped down the effects of the wave shocks.

All ships must be fitted with the equivalent of two independent steering gears and in the smaller vessels this was effected by two large hand-operated wheels suitably geared to operate direct over the rudder stock. This was superseded by the use of tackles with the hauling parts taken to steam mooring winches so that the vessel could be steered by power from aft.

Telemotor Gear.—This gear is designed to do away with the system of rods and chains leading along the deck. It also allows the steering engine to be placed alongside the quadrant, thereby giving direct action. Helm movements are controlled on the bridge by what may be described as a pressure machine. To this machine the wheel is attached, and it is connected to the steering engine by two small pipes which are below decks and protected from damage. Wheel movements cause oil pressure to act along the pipes and operate the valve of the steering engine, which in turn acts directly on the quadrant. Fig. 12.

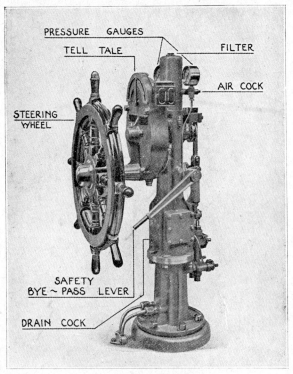

PRESSURE GAUGES

TELL TALE

FILTER

AIR COCK

STEERING
WHEEL

SAFETY
BYE ~ PASS LEVER

DRAIN COCK

FIG. 12.—Telemotor Steering Gear.

Hydraulic Steering Gear.—Most steering gear today is of the hydraulic type whereby four fore-and-aft diagonally connected rams operate in pairs either side of a crosshead attached to the rudder stock. By means of a variable-delivery pump of the Hele-Shaw type, oil can be transferred from one ram or set of rams to the other and by so doing exerts a turning moment on the crosshead which moves the rudder. The variable-delivery pump is pushed or pulled on stroke by the telemotor gear according to the movement of the wheel on the bridge and the movement of the crosshead operates the hunting gear returning the pump to the neutral position pending further movements of the wheel. In the event of heavy seas striking the rudder, there is a by-pass valve which allows the rams to give to the blow and the hunting gear restores the position immediately afterwards. There is an emergency arrangement for steering from aft by direct connection to the pump controls and a second independent pumping system covers requirements for secondary means of steering.

The pumps may be driven by high-speed forced lubricated steam-engines in which case it is called "Steam Hydraulic Steering Gear" or by electric motors in which case it is called "Electro-Hydraulic Steering Gear".

All Electric Steering Gear.—In this case the connection between the wheel on the bridge and the rudder motor which is coupled through gearing to the rudder stock, is carried out electrically by means of rheostats at each end. There is a continuously driven electric motor taking its supply from the ship's mains and driving a generator and exciter. The armature of the generator is wired to the armature of the rudder motor and the field of the generator is fed from the exciter. This field only becomes active when a movement of the wheel on the bridge upsets the balance of the resistances as between the bridge and aft and when this happens the generator supplies current to the rudder motor which moves the rudder in a direction depending upon which way the wheel is turned. The movement of the armature also moves a slider on the rudder rheostat to restore balance with that on the bridge when all movement stops pending further movements of the wheel

Steering Flat.—Compartment situated above the rudder. It contains the steering engine which is attached to the quadrant on the rudder post.

Stokehold.—Space where the boilers are laid, and where the stoking or firing is done.

Stokehold Plates.—Iron plates which form the flooring of the stokehold.

Storm Plates.—Large iron plates hinged to the fiddley top, against each of the gratings, and over which they are fitted in bad weather, to prevent water finding its way to the top of the boilers or into the stokehold.

Strong Room.—Extra strong storage place for valuable cargoes in mail and passenger steamers.

Strum Box.—*See* Rose Box.

Table.—A construction on deck in the shape of a table through which the foot of the mast passes and providing a means of supporting the goosenecks for the heels of the derricks. Fig. 10.

Taffrail.—Rail fitted to the after end of the poop.

Tail End Shaft.—The short end of propeller shaft which lies in the stern tube. When the propeller is to be unshipped, the tail end must be drawn inside the tunnel to allow the propeller to swing clear, therefore, they are made short for easy handling.

Telegraphs.—An automatic means of communicating orders between the navigating bridge and engine-room. Docking telegraphs, situated on the forecastle head and poop, and connected to the bridge, are used for the same purpose.

Telephones.—Communication with the bridge from the forecastle head, poop, engine-room and crow's nest is effected by telephone in most vessels.

Tonnage Hatch.—Hatchway and space which was only provided to conform to certain regulations. It was only found in Shelter Deck ships, and was always situated on the after deck. Wooden hatch covers fitted over the opening, and although tarpaulins were supplied, there were no cleats for battening down. In theory, water could enter the hatch and scuppers were provided to clear it overside. There is no longer any advantage in constructing a tonnage hatch.

Tonnage Mark.—A triangle and horizontal line on the ships side indicate that certain spaces need not be counted in the gross tonnage, if the mark is clear of the water. If it is submerged the larger of the two tonnages applies. This is the modern equivalent of the shelter deck ship.

Transom.—The wide part of a stern frame above the propeller. It is exactly the same shape, and in the same position as the wide stern of a jolly boat.

Truck.—Extreme top of the mast, which takes the form of an oval button, fitted with small sheaves through which flag halliards are rove.

Tunnel.—The space formed by the casing over the propeller shaft. It runs through the holds from the engine to the stern tube.

Tunnel Escape.—A means of escape from the tunnel. It is situated right aft, with a door opening on any of the upper decks.

Ventilation.—The long funnel and large cowl ventilators were the mark of the natural draught coal burning ship. Forced draught of varying types was introduced until the enclosed engineroom ventilated by fans is usual today. Hold ventilation is much more carefully controlled than ever and except for certain bulk trades the ventilation will be by fans and it will be controlled according to the dew point of the air outside and the temperature inside the hold.

"Waist" of the Ship.—All the middle section between the forecastle head and poop is known as the "waist".

"Waist" Pipes.—These are for the same purpose as shoulder and quarter pipes, but are situated in the "waist", at the opposite ends of the fore and after well decks.

Wash Ports (or Freeing Ports).—Small square doors hinged to the bulwark and hanging free, and which help to rid the deck of water during bad weather. A weight of water on the inside opens them, but a sea striking the outside closes them.

Waterways.—Gutters at the ship's side which carry water to the scupper holes and pipes.

Wheelhouse.—A house situated centrally on the navigating bridge. Originally it protected the man at the wheel while the officer of the watch stood in the open or in one of the wing cabs. In modern ships it has been extended to include most of the bridge and the chartroom as well.

Whelps.—Longitudinal strips on the warping end of a winch which help the rope to grip the rounded surface.

Whistles.—One, two, or even three whistles will be found on the forward side of the funnel. They are connected to, and worked from, the bridge either by means of whistle lanyards or electricity. Compressed air is used for whistles on motor-vessels.

See Annex III. Regulations for prevention of collisions at sea.

DECK MACHINERY.

Steam Winches.—Individual winches vary so much that no one can be sure just how they will act until a trial has been made. For this reason it is necessary to be extremely careful when, after heaving a load up, it is about to be lowered.

With a very heavy load it is sometimes only necessary to completely shut the steam off and the winch will "walk back", in which case it is only necessary to turn it on again to stop the winch. Fig. 13.

Fig. 13.—A Totally Enclosed Splash Lubricated Steam Winch.

In other cases the winch lever must be reversed or put in the "lower" position before anything happens, then to stop the winch the lever is returned to the original "heave up" position, with perhaps a small touch of steam to help.

At all times a winch driver must keep his wits about him, but never more so than when about to "lower away" with a strange winch.

Winches.—According to the size of the ship and the trade she is engaged in, either one, two, or even four winches will be found at each hatchway for heaving cargo in and out of the holds. Most vessels have a special mooring winch or capstan right aft for heaving in the mooring lines.

Before using a winch, any water lying in the pipes, through condensation of steam, must be run off. This is done by opening the drain cocks underneath the cylinders. If a little steam is turned on, it will help to blow the water through, especially if the reversing lever is moved backwards and forwards a few times. A winch will not work smoothly until all the water has been got rid of.

Out of Gear.—When the pistons and cranks are moving but the centre and end drums do not revolve the winch is out of gear.

Single Gear.—Underneath the centre drum there will be found two levers which control the gear clutches on the driving shafts. Usually, the smaller of the two controls one clutch, and when it is engaged the winch is in single gear.

Double Gear.—As a rule the larger lever operates two clutches. These are connected by a bar which prevents both of them engaging at the same time. If the bottom one is put in, the top one automatically comes out, and vice versa. When the top clutch is engaged along with that of the single, the winch is then in double gear.

Safe Clutches.—In the older type of open winch it was usual to take the precaution of preventing the clutches from becoming disengaged particularly when taking a heavy weight aboard, by cutting four pieces of wood to fit in between the clutches and lashing them to the driving shaft.

Electric Winches.—Electric winches have now been developed to a stage of complete safety and are automatically safeguarded against damage as a result of mishandling the controls. The type shown in figure 14 is operated by a controller and a footbrake directly mounted on the winch, but they are more often operated from a remote control position. Incorporated in all these winches is a magnetic brake of the disc type with large friction surfaces which is automatically released as the power is switched on by means of the controller and comes into operation again as the power is switched off so that the load is held. It cannot "run back" or "walk back" as could be the case with the older steam winch.

In the latest types of electric winch the speed of hoisting or lowering is completely controlled electrically and no foot-brakes are fitted. All mechanical braking is carried out by the magnetic

FIG. 14.—Electric Winch.

FIG. 15.—Windlass.

brake and the winch automatically stops when the load touches down. With only one control it is possible for one operator to control two winches by grouping the remote controllers in pairs.

Windlass.—This is a special type of steam or electric winch which is used to heave the anchor up from the sea-bottom and controls the cable when running out.

When not employed in anchor work, the windlass can be used for heaving in mooring lines. The brake is screwed up tight and the gypsy, which carries the cable, is put out of gear by turning the small wheel fitted for the purpose. (In the older type of vessel an iron lever was used for this operation.) Once the gypsy is out of gear, the drum ends of the windlass can be used for warping purposes but as the anchor is then "on the brake" the anchor cable should be held for safety in the bow stopper or by devil's claws or chain stopper. These must be cleared before letting go the anchor by releasing the brake.

When ready to heave the anchor up, the gypsy is first put into gear by operating the small control wheel and the brake is then slacked off.

Most windlasses today are driven by electric motors which permit conveniently placed remote controls, and in some ships the anchors can be let go or hove up from the bridge.

Mooring Capstan.—In many ships the old-fashioned type of warping winch has been replaced by the mooring capstan in which the barrel is vertically mounted and driven by an electric motor, or by steam power, through a worm and bevel gearing. They have the advantage that the motor or engine can be situated remote from the capstan and this leaves the deck space clear.

Self-tensioning Winches.—Self-tensioning mooring winches have recently been developed to meet the needs of the large oil-tankers and bulk carriers where the rapid rate of loading and discharging is such that the elevation of the vessel in relation to the wharf or dock is continually changing. This in the ordinary way would involve a constant watch being maintained to ease off or take in the slack of the mooring wires. When automatic tensioning winches are used the mooring wire is permanently secured and stowed on the barrel and adjustments of the mooring line are dealt with automatically. This makes the operation of mooring much more simple and safe and is a great saver of labour. The winches can be hand operated.

Bow Stopper.—An appliance on the forepart of the windlass, over which the anchor cable runs. It is designed to secure the cable when the windlass brake is slackened off.

By lifting a lever, the weight of a movable chock is held while a pin is withdrawn from underneath. When the lever is lowered again, the chock drops a few inches, and allows the rounded part of a horizontal cable link to catch in the rounded part of the compressor thereby preventing the chain from running out. The pin previously taken from underneath is now fitted over the top of the link, and this prevents the cable from accidentally jumping out of the chock. To avoid the friction and sharp angle as the cable enters the hawse pipe, the cable may lead over a substantial pulley which may also incorporate a special stopper device.

Wire Reels.—Mooring wires when not in use are wound on wire reels fixed to the deck in convenient positions. The wire must be taken off the reel and flaked down on deck before before paying it out.

CHAPTER III.

MANUAL SEAMANSHIP

Rope Making Materials. Splices, Knots, Bends and Hitches. Wire Rope—its Manipulation, Splicing, Worming, Parcelling and Serving. Grades of Canvas, Sewing. Boatswain's Chair. Moorings. Sailorising Jobs.

ROPE MAKING MATERIALS.

Rope.—It is known that ropes were used in China at a very remote period, and that the Egyptians made them from papyrus and palm fibres, as well as hides, but there is no record of who first conceived the idea of making rope.

At the present day ropes are made from either vegetable fibre, man made fibre or metallic wires. The principal fibres used are as under.

Manila.—Abaca is the fibre which is usually referred to as manila, but it is more commonly known by the name of the port in the Philippines from which it is exported, and where it grows almost exclusively.

Being strong and durable it makes excellent rope, and as it does not rot it is never tarred. It does, however, swell considerably when wet, so is not altogether suitable for running rigging. It stretches 20 per cent to 30 per cent.

Hemp.—Although hemp was once imported from European countries in large quantities, and used almost exclusively in the manufacture of rope, it is now almost completely superseded by manila, with the result that hemp is now only found in boltrope, and a few of the smaller classes of lines, most of which are of the tarred variety. It does not swell when wet, and for this reason makes good running rigging.

Coir.—A fibre which is not so durable as hemp. It comes from the coco-nut palm and will soon rot if stowed away wet.

Although about one-third lighter it is only one-quarter the strength of hemp rope, and as it floats very lightly on the water it is very useful as a warp. It is usually employed for all purposes where a rope with a good "spring" is required, such as in a towing spring. About 25-30 fathoms may be attached to a towing wire to act as a "spring". It stretches from 40 per cent. to 50 per cent.

Coir rope is sometimes referred to as "bass" or "grass rope".

Cotton.—Rope composed of cotton fibre is mostly found on yachts, and is used to a greater extent in America than elsewhere.

Sisal.—A material which comes from the fibre of the leaves of the American aloe. It is very white in colour, almost as strong as hemp and stands up to sea-water quite well.

Nylon.—Nylon is one of the popular modern synthetic fibres used to form rope which amongst other things has the advantage of combining lightness with strength and extreme flexibility. It is very durable, easy to handle and can withstand shock loads but like other synthetic fibres it is somewhat costly to manufacture.

Polyester, Polypropylene, Polythene.—Man made fibres, the last two have the advantage that they will float but none of the three are as strong size for size as nylon.

Formation of Rope.—The formation of all rope depends on "twist". Fibre is twisted up into yarns which are then twisted up into strands. These in turn are laid up or twisted up to form the complete rope. Friction, and the inclination of each strand to unlay holds the other in place.

Yarns.—These are composed of threads of fibre well and evenly spun into what is known as a standard 24-thread yarn. The size of this yarn is such that if 24 such yarns are formed into a strand and three such strands are laid up to form a rope, the rope will measure three inches in circumference.

Strands.—The number of yarns in a strand depends on the size of the rope, and each strand is composed of yarns laid up right-handed in the case of a left-handed rope, and left-handed for a right-handed rope.

Ropes may be composed of either three or four strands, but three is the most common in use.

Lay of the Rope.—Ropes are either *right-hand lay* or *left-hand lay*, according to the direction in which they are finally "laid-up" or "twisted up". With the former the strands run from left to right, and with the latter from right to left. Practically all ropes are *right-hand laid*.

If a rope is "twisted up" very tight and hard it is said to have a *hard, firm or short lay*. One effect of this is that pliability and breaking strain are reduced, but the rope is not so liable to absorb water and lose its shape.

On the other hand, a rope which is loosely "twisted up" or "laid up" is said to be of *soft* or *long lay*. This has the effect of increasing the pliability and breaking strain, but soft laid rope is more liable to absorb water and lose its shape.

When a rope has only sufficient "twist" to combine pliability, strength and ability to withstand hard working conditions, it is said to be of *standard* or *plain lay*, and this is the most common form in use.

TYPES OF ROPE

Hawser-Laid Rope.—This is the commonest form of lay. It is the ordinary three-strand rope laid up right-handed, which is used for practically all purposes.

Shroud-Laid Ropes.—Four-strand ropes laid up right-handed around a central "heart".

Cable-Laid Rope.—A left-handed rope, the difference in which is very noticeable. Each of the three strands is a complete rope in itself, consisting of three strands laid up right-handed; it would therefore be more correct to call it nine-stranded. The sizes range from about 5 inches to 18 inches.

Cable-laid ropes are principally employed as towing "springs". For this purpose they are spliced to a length of wire, but it is essential that the lay of the wire and that of the rope is the same. As the lay of a wire is usually right-hand, cable-laid ropes intended for this purpose are always made with a right-hand lay.

Square Line.—As the title implies the cross section of this type of rope is square in that it is made up of four pairs of strands, two pairs laid up left-handed and two pairs laid up right-handed. The resulting rope is tough, durable, flexible and does not kink. It makes an excellent mooring-rope and when taken round the drum-end of a winch there is no tendency for the turns to ride.

The manufacturers supply pamphlets on the special procedure when splicing this type of rope and these instructions should always be consulted.

Water-Laid Rope.—Same as cable-laid.

Warp Lay.—The only difference between this type and the cable laid rope (three three-strand ropes laid up together) is that the first and final lays are very hard. Sometimes four-strand warps (containing 12 strands) are supplied, and for special types of work a five-strand warp (containing 15 strands) is sometimes used.

Unkinkable Lay.—This type of rope is specially made for life-boat falls. Individual yarns were spun the same way as the strands. This was done to eliminate the tendency of the whole tackle to spiral when the boat touched the water and the weight came off. Once spiralled the falls were difficult or impossible to recover.

Preserving Ropes.—Ropes which may be constantly immersed in water are often "tanned", or "barked" by the application of "cutch", to preserve them. Archangel tar and coal tar are also used for the same purpose, but this reduces the breaking strain by about 10 to 15 per cent. in the case of hard fibre rope, and about 30 to 40 per cent. in the case of soft fibre ropes. It also stiffens the rope and makes it heavier.

During manufacture a small proportion of lubricant is added to soften and lubricate the fibres. All ropes are "oil spun" in this way, except those which are made for some special purpose where oil is likely to cause a stain. Ropes without oil are said to be Dry Spun.

Small Stuff.—Apart from the heavy ropes found on all vessels, there are a number of light lines and "cordage" more generally known as "small stuff". They are as follows:—

Boltrope.—A good quality three-strand right-handed hemp rope used for sewing to the edges of sails for the purpose of strengthening them. Boltrope is made in sizes ranging from half-inch to six inches circumference or 4mm to 48mm diameter.

It is soft laid and well stretched before being made into coils, so as to render it soft and pliable.

Only the small sizes of boltrope are ever found on board the average steamer—hence the reason for including it among "small stuff".

Point Line.—Small size three-strand manila rope, ranging from about 1 inch to $1\frac{3}{4}$ inch circumference, 8mm to 14mm diameter. Sizes are always designated by the number of "threads" of which 15, 18 and 21 are the popular numbers.

"Ratline".—A three-stranded tarred soft hemp rope supplied in coils of 120 fathoms (220m). It is measured by the number of threads or ropeyarns it contains. 9 thread is about 8mm diameter and 24 thread is about 14mm diameter.

Heaving Lines, Making Up.—Always make a small coil, so that the line will be ready for throwing if required in a hurry.

A coil finished off by making two half-hitches with the end around all parts is not always handy when casting adrift again.

The best way is to proceed as follows:—

Form a small bight about 3 feet from the end and pass through the coil. Twist this bight over once, so that the short end is over the standing part, and so forms a sort of "eye". Pass the end around all parts and through the "eye". When pulled tight, this hitch will hold a line firmly in position and is very easy to let go again.

Loglines.—Plaited hemp or polythene lines supplied for use with the patent log, in coils of 75, 110 or 220m.

Flag Halliards.—Usually a four stranded hemp line but it may be plaited flax or polythene about 10mm diameter. Supplied in coils up to 220m or hanks of 55m.

Hand Leadlines.—These are three-stranded dressed hemp lines of 10 mm diameter and are water- or cable-laid (left-handed), being supplied in 55m hanks.

Deep Sea Leadlines.—Dressed hemp lines of 11mm diameter containing three strands water- or cable-laid (left-handed) and supplied in 220m coils.

"Boat Lacing".—High-grade three strand dressed hemp lines supplied in 55 metre hanks and used for lanyard and lacings.

Hambro'-line.—Three-yarn or three-stranded right-handed tarred soft hemp made in two sizes, 3- or 6-thread. It is usually supplied in 30-fathom hanks.

Houseline.—Three-yarn tarred hemp laid up left-handed.

Marline.—Two-yarn tarred hemp laid up left-handed and is usually supplied in 1kg. balls.

Spunyarn.—Two or more yarns tarred hemp usually laid up loosely right-handed and largely used for "serving" ropes or splices. It is supplied in 3kg balls.

Samson Lines.—Very light hemp lines sold in 55 metre hanks. They are similar in appearance to fishing lines, but are seldom seen in the Merchant Service.

Seaming Twine.—A three-ply twine spun from the best flax, and made up in hanks of 1 lb.

Roping Twine.—This is also sold in 1-lb. hanks and is usually five-ply.

Machine Twine.—Is made up in balls or cops weighing 1 lb., and is usually chemically treated to resist rot and mildew. It is prepared from a high-grade flax and may be either two- or three-ply.

Strength of Rope.—To find the ultimate strength of hemp, manila or vegetable fibre in general, square the circumference of the rope and divide by three. This will give the breaking-strength in tons and the safe working load would be taken at one-sixth of that value. For example, take a three-inch rope, this would give us 3 tons breaking-strength with a safe working load of $\frac{1}{2}$ ton. In practice one would use the appropriate tables provided by the manufacturers.

With the change over to the metric system the size of all natural and synthetic fibre rope and steel wire rope are measured by their diameter in millimetres and the rough formulae indicated in the previous paragraph to obtain the breaking stresses for rope will no longer apply. Chain on the other hand has always been measured by the diameter of the metal forming the link and with the conversion to the measurement being in millimetres, it has been found possible to reduce the formulae to a similar type for all materials.

In the absence of information from the Manufacturers, the following formulae will give reasonable indication of the breaking stress whilst the safety factor for the Safe Working Load is still to be taken as one-sixth of the Breaking Stress.

METRIC FORMULAE FOR BREAKING STRESSES OF NATURAL AND SYNTHETIC FIBRE ROPE, STEEL WIRE ROPE AND CHAIN

MATERIAL		*FACTOR*
FIBRE ROPE, 3-STRAND (hawser laid).		
Grade 1 Manila	7mm to 144mm ⎫	
High Grade Manila	7mm to 144mm ⎭	$2D^2/300$
Polythene	4mm to 72mm ⎫	
Polypropylene	7mm to 80mm ⎭	$3D^2/300$
Polyester (Terylene)	4mm to 96mm	$4D^2/300$
Polyamide (Nylon)	4mm to 96mm	$5D^2/300$
FLEXIBLE STEEL WIRE ROPE.		
6×12	4mm to 48mm	$15D^2/500$
6×24	8mm to 56mm	$20D^2/500$
6×37	8mm to 56mm	$21D^2/500$
STUD LINK CHAIN.		
Grade 1	12·5mm to 120mm	$20D^2/600$
Grade 2	12·5mm to 120mm	$30D^2/600$
Grade 3	12·5mm to 120mm	$43D^2/600$
OPEN LINK CHAIN		
Grade 1	12·5mm to 50mm	$20D^2/600$
Grade 2	12·5mm to 50mm	$30D^2/600$

The diameter D is expressed in millimetres, the breaking stress in tonnes.

Lifting Power of Tackle.—If the weight that a single part of rope is capable of suspending is multiplied by the number of parts at the movable block, it will give the lifting power of a tackle but one-fourth must be subtracted to allow for frictional resistance.

Opening a Coil.—Very often the label attached to a coil of rope contains instructions as to how it should be opened up, but if not, proceed as follows (Figs. 1, 2):—

FIG. 1.
Opening a
Coil (from
outside).

FIG. 2.
Opening a
Coil (from
inside).

(a) *Ropes with a Right-hand Lay.*—Strip bagging, find outside end and lay coil on whichever side will permit the end to come away left-handed (anti-clockwise), i.e. allow the coil to rotate as on a turntable.

Another method is to leave the bagging intact, and lay the coil so that the inside end will come away left-handed, i.e. the inside end should be at the bottom and taken up through the centre of the coil.

(b) *Ropes with a Left-hand Lay.*—Either the inside or outside end may be used, but the rope must come away from the coil right-handed (with the hands of the clock).

Coiling Rope.—Ropes laid up right-handed must be coiled down in a clockwise direction, and left-handed ropes in an anti-clockwise direction.

To Thoroughfoot a Rope.—If through a mistake or any other cause a rope is taken off a coil the wrong way, it will become full of turns and will be required to be thoroughfooted in order to get rid of the turns or kinks. To do this it must be coiled down large left-handed and the bottom end passed up through the centre of the coil and again coiled left-handed. Repeat the operation two or three times and finally bring the end up through the centre and coil right-handed.

Stretching Rope.—Very often a rope requires stretching before it can be used in making or reeving off new gear. A half-hearted stretch by hand or steam winch may be good enough in some cases, but in others a more effective process requiring time will be necessary.

A good method is to make one end fast in some convenient place and heave on the other end with a tackle at frequent intervals. When the rope sags, the slack can be taken in on the tackle and if left lying all day and all night, it will be ready for use on the following morning.

SPLICING.

Rope Eye Splice.—Unlay enough end to make about three tucks (one turn for each tuck to be made), then form an eye with the ends on top. That is, the three ends must be running diagonally across the rope from right to left, on top. Fig. 3.

Take the middle end (1) and tuck it underneath the nearest strand on the standing part from right to left. Pick up end (2) and take it over the strand which has the middle end (1) under it, and tuck it under the next one on the left. Turn the whole job over, then take end (3) over to the right, then back to the left and tuck it under the only strand in the standing part which has nothing under it. If all this has been done correctly there should be an end coming out between each of the strands. Should two ends come out between the same strands the splice is wrong. For the next round of tucks each end is led over one strand and under the next, towards the left.

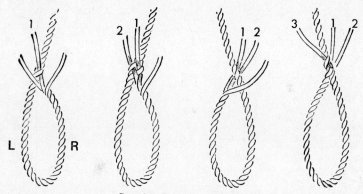

FIG 3.—Rope Eye Splice.

As a general rule, two full rounds of tucks are considered sufficient for ordinary purposes, and three when the rope has to bear any strain.

For neatness, a splice may be tapered by adding an extra round of tucks with halved strands. These should be cut on the underneath side, so that the short or cut ends are hidden.

Dogging the Ends.—After putting an eye splice in a mooring rope, the usual practice is to "dog the ends" before cutting them off. Each strand is halved and unlaid, then each pair of adjoining halves are whipped together. Fig. 4.

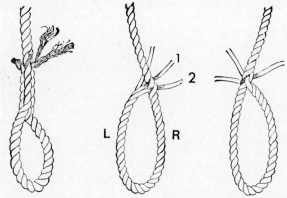

FIG. 4.—" Dogging the Ends".   FIG. 5.—Sailmaker's Eye Splice.

Pick any pair, then if the ends are long enough, choose a good yarn from each half strand at the point where they nearly touch.

Pass these yarns around the two halved strands in opposite directions and make an overhand knot each time they meet.

Continue doing this until a good whipping has been passed, then finish off with a reef-knot.

If the yarns belonging to the strands are not long enough, any long yarn can be used.

Sailmaker's Eye Splice.—This is a very simple splice which is used by sailmakers because it looks neater on roping sewed to canvas. It is never used on any rope which has to bear a strain as it is likely to "draw".

With the three working ends crossing the standing part from left to right take the middle end (1) and tuck it under the nearest strand from left to right, tuck end (2) under the next strand on the right again from left to right then turn the job over and tuck end (3) under the remaining strand from left to right.

The splice is continued by spiralling each end around the strand it was first tucked under. If the working ends are progressively tapered the splice will run away to nothing and when sewn to a sail the result will be very handsome. Fig. 5.

German Eye Splice.—Place ends in the same manner as for an ordinary eye splice, then tuck the middle end through nearest strand towards the right. Next, tuck the right hand end under the same strand, but towards the left. This means that the first two ends are crossed under the same strand. The third end is tucked under the left hand strand, towards left, in the ordinary way.

Succeeding tucks are put in over one and under one in the usual way.

Short Splice.—The general rule for a short splice is to unlay one turn on each end for each tuck to be made, but it is as well for the novice to make it one extra and be on the safe side.

Place the two unlaid ropes together, so that the ends on the right-hand side lay in between the ends on the left.

With a large-sized rope a firm seizing may be passed around the exact meeting place of the two ropes, but with lighter ones it is just held firmly in the hand while the tucks are made. When tucking, each end is simply passed over one and under one, and about two tucks each way is the usual number made.

A loose splice is of little use, therefore all ends must be pulled as tight as possible.

An easy method (not always permitted) of holding the two ropes together is to form an overhand knot with each pair of ends which come together from opposite sides, taking care that the ends follow the lay of the rope and not across it. Take one in each hand and pass one over the other in the same way as the first part of a reef-knot and pull tight. Do the same with the other two pairs, then begin the tucks. Fig. 6.

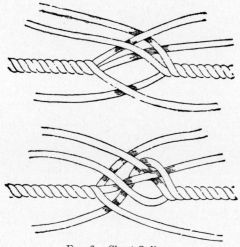

FIG. 6.—Short Splice.

Chain Splice.—The rope tail spliced to a chain stopper is usually of such a size that it is too thick to pass through the link of the chain, in which case it must be chain-spliced.

Unlay about 18 inches of one strand, pass the other two through the link and form an eye about 1½ inch long, leaving about 8 inches of end for splicing.

Separate these two ends. Take whichever will lie neatest and lay it up in the vacant score until it meets the first end which was unlaid. Finish off these two ends in the same way as for a long

FIG. 7.—Chain Splice.

splice (overhand knot with the lay of the rope, and each end halved and tucked twice, over one and under one). The remaining strand is also tucked over one and under one. Fig. 7.

Long Splice.—In actual practice long splicing is very seldom resorted to. It is a wasteful method, and ropes requiring to be joined in this manner are usually replaced with new ones.

A good deal more end has to be unlaid for a long splice, but the actual amount is dependent upon the size of rope to be joined, the length which can be spared for the purpose, and the weight it has to bear when in use.

As a general rule, about three or four times the amount required for a short splice is unlaid, and the ends are placed together in exactly the same manner as for that method.

When this has been done proceed as follows:—

Pick out any two ends which cross one another from opposite sides, unlay one of them a good distance, and lay the opposite number in its place until only a few inches are left. Cut off surplus from unlaid end, and do exactly the same with a second pair, but work towards the opposite direction.

The third pair are left in their original position. Fig. 8.

This gives us three pairs of ends an equal distance apart, and to dispose of these an overhand knot is made with each pair, but the ends must follow the lay of the rope and not across it. After pulling tight, divide each strand in two and tuck one half of all strands two or three times, over one and under one.

If possible, the rope should be well stretched before cutting the ends off close.

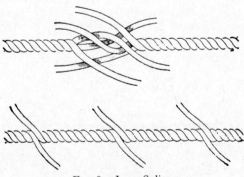

Fig. 8.—Long Splice.

Some seamen also put a palm and needle whipping at each of the three points where the ends are knotted and tucked, to make sure they will not come adrift through wear and tear.

Splicing Four-Strand Rope.—Divide the four working strands so that they are two on each side of the standing part. If you stand at the eye with the standing part away from you, then the two strands on your left are going against the lay. These two strands go in under the first available strand, and the first one goes under two strands while the second only goes under one. The third strand goes under the next strand on the right from right to left and the fourth strand goes under the remaining strand on the right, again from right to left. This gives one strand coming out under each strand of the standing part. The splice is continued by tucking each strand over one and under one to the left for a further three rounds.

Splicing Eight-Strand Square Line.—Although this is eight-stranded rope, it is not an eight-strand sennit. It is a four-strand sennit made with four pairs of strands. These pairs can be identified as two pairs with left-hand threads and two with right-hand threads. Having identified them, whip them together at the ends and reduce the chance of error.

Look at the lay and decide how much working end you will require for, say, four tucks, and put on a good whipping before unlaying the pairs.

Form an eye so that two R.H. pairs lie down one side of the square and two L.H. pairs lie down the opposite side. The rule is that like follows like, so tuck one R.H. pair under a L.H. pair on the side of the square then tuck the next R.H. pair under the next L.H. pair, still on the same side of the square. On the opposite side the L.H. pairs follow the L.H. strands under the R.H. pairs. You now have tucks on two sides but no tucks on the top or bottom. All the ends will now be on the bottom, and the next tuck sees each pair crossing the bottom still following like with like. In similar fashion, the third tucks now come up the sides so that the fourth tucks will cross over on the top. At this stage the whippings can be cut and another tuck put in, using one of each pair only.

Cut Splice.—This is often used for joining two wires instead of using a long splice, but it is a very unsightly and lumpy join.

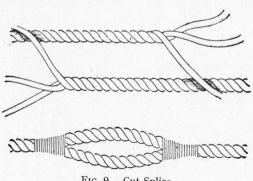

FIG. 9.—Cut Splice.

Unlay the two ends as for an eye splice, overlap the standing parts a few inches, and tuck all ends in the same manner as for an eye splice. Fig. 9. If desired, an eye can be formed by increasing the overlapping to any required distance.

Splicing Plaited Line.—To splice an eye or make a short splice with plaited line is a tedious business requiring patience, and is seldom done at sea.

The usual practice is to unlay the end some two or three inches, form the eye, lay the ends along the standing part and serve over tightly with strong twine.

Stropping Baskets.—Baskets must be fitted with strops before they can be used for hoisting purposes. This may be done in several ways, but the simplest method is to make an ordinary strop, and instead of joining the ends with a short splice, make a cut splice which will be big enough to be seized around the bottom of the basket. Strong seizings must also be put on the strop running up each side of the basket. Spun-yarn may be used for seizings, but seizing wire is better.

Very often the lip of the basket is served over all round the edge with a full strand of yarns, and these may be slightly spaced or closely served as desired.

KNOTS, BENDS, AND HITCHES.

Knots with Single End

Overhand Knot
Figure-of-Eight Knot
Bowline
Half Hitch
Clove Hitch
Cow Hitch
Awning Hitch
Rolling Hitch
Timber Hitch
Marline-spike Hitch
Blackwall Hitch
Double Blackwall Hitch

Knots with Own Strands

Wall
Crown
Manrope Knot
Single Matthew Walker
Double Matthew Walker

Knots and Bends Uniting Ropes

Reef Knot
Sheet Bend
Single Carrick Bend
Double Carrick Bend
Common Whipping
Palm and Needle Whipping

See Figs. 10 to 34.

FIG. 10.—Overhand Knot.

FIG. 11.—Figure-of-Eight Knot.

FIG. 12.—Bowline.

 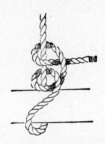 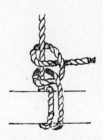

FIG. 13.—Half Hitches

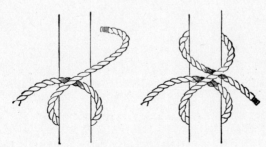

FIG. 14.—Clove Hitch.

C

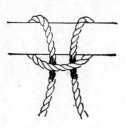

Fig. 15.—Cow Hitch.

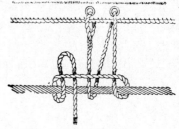

Fig. 16.—Awning Hitch.

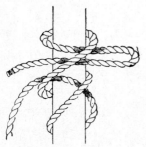

Fig. 17.—Rolling Hitch.

Fig. 18.—Timber Hitch.

Fig. 19.—Marline-spike Hitch.

Fig. 20.—Marline Hitch.

Fig. 21.—Blackwall Hitch.

Fig. 22.—Double Blackwall Hitch.

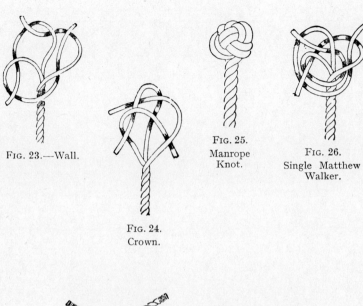

Fig. 23.—Wall.

Fig. 24.
Crown.

Fig. 25.
Manrope
Knot.

Fig. 26.
Single Matthew
Walker.

Fig. 27.—Reef-knot.

Fig. 28.—Sheet Bend.

Fig. 29.—Double Sheet Bend.

Fig. 30.—Single Carrick Bend.

FIG. 31.—Double Carrick Bend.

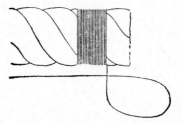

FIG. 32.—Common Whipping.

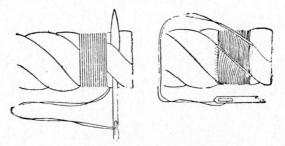

FIG. 33.—Palm and Needle Whipping.

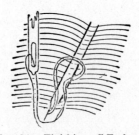

FIG. 34.—Finishing off End.

Heaving Line Knot.—Form a bight about 5 feet from the end of the line. About 8 inches from the actual bend start wrapping the

spare end around both parts of the bight, working towards the bend
itself, and make about 10 turns. Pass the remainder of the spare
end through the small bight which is left, and pull tight on the
standing part of the line. This leaves a good heavy knot which
will carry the line well when throwing. Fig. 35.

FIG. 35.—Heaving Line Knot.

Monkey's Fist.—This is a good heavy knot which is formed on
the end of a heaving line to give carrying power to the end when
throwing.

About 9 feet from the end, make three small coils of line over the
hand, about 4 inches in diameter. Then make three more turns
around or across the middle of the first three. Pass another three

FIG. 36.—Monkey's Fist.

FIG. 37.—Joining Heaving Line Quickly.

turns around or across the second three, but inside both ends of the
first three. If correct, the remaining end will come out alongside the
standing part, and to this it is eventually spliced, when all parts are
tightened up. Before tightening up, poke a piece of waste or
oakum into the centre of the ball to act as a heart. Fig. 36.

WIRE ROPE.

Wire Rope.—Prior to 1874 wire ropes were not sufficiently pliable for marine purposes, but the introduction of flexibility has now enormously increased their sphere of usefulness. At the present time wire ropes have superseded both hemp and manila for practically all purposes on board ship.

Flexibility was attained by increasing the number of wires used in the formation of the rope, and as the purposes for which the rope is intended also determined the number of wires in its make-up, it will be found that ropes intended for different types of work are of different construction.

These types are many and varied, but the most popular rope for all marine purposes is the 6/19. That is to say, it has six strands with 19 wires in each strand.

Each wire is constructed so that the strands are laid around a hemp heart, and in some makes a heart will be found in each strand also.

Construction, Size and Use.—The following table will give an idea of the construction sizes and uses of the wire ropes likely to be met with on board ship.

	Construction	Diam.	Names
(1) For Standing Rigging Funnel Guys, Stays, etc.	6/7 and 7/19	12mm to 48mm	Non Flexible
(2) For Cargo Falls, Hawsers, Towlines, etc.	6/12 and 6/19	8mm to 24mm	Ordinary Flexible
(3) For Cargo Falls, Hawsers, Towlines, etc.	6/24	8mm to 24mm	Extra Flexible
(4) For Hawsers, Towlines, etc.	6/37	20mm to 72mm	Special Extra Flexible

Preformed Wire Rope.—A new type of wire rope the special feature of which is that the wires and strands, before laying up, are shaped to the spiral form which will be required in the finished rope.

This counteracts the tendency of all wires to "spring" apart when cut or unlaid.

Steel.—For all general purposes, the steel used in the manufacture of wire ranges from mild steel for non-flexible wires to hard drawn plough steel for the flexible type.

Breaking Load of Flexible Wire.—To find the approximate breaking load in tons, square the circumference in inches and multiply by 2.

SIX-STRAND WIRE ROPE

APPROXIMATE BREAKING LOAD

Circum-ference	Diameter	TONNES			
Inches	Milimetres	6×12	6×24	6×36	6×37
1	8	1·94	2·60	—	—
$1\frac{1}{4}$	10	3·02	4·06	—	—
$1\frac{1}{2}$	12	4·35	5·85	8·55	—
$1\frac{3}{4}$	14	5·93	7·96	11·60	—
2	16	7·74	10·40	15·20	—
$2\frac{1}{2}$	20	—	16·20	23·80	—
3	24	—	23·40	34·20	—
$3\frac{1}{2}$	28	—	31·80	46·60	—
4	32	—	41·60	60·80	—
5	40	—	65·00	95·00	68·4
7	56	—	—	—	134·0
9	72	—	—	—	222·0

Compound Wires.—Wires covered with manila yarns are not common in the Merchant Service, except in the case of cable steamers where a special type is used for anchoring buoys (in conjunction with chain), or for attaching to grapnels.

Grapnel and Buoy Ropes.—These are always manila-covered wires, and are used principally for heaving telegraph cable up from the sea bed for the purpose of repairs, in depths of from two to three or even more miles.

They are made from the best Homo steel and no joints are allowed in their construction—either welded or otherwise.

Two sizes are in general use as under—

Three by Three.—Three strands with three wires in each, covered individually with five manila yarns laid up right-handed.

The three wires are laid up together left-handed to form a strand and the three strands are laid up right-handed.

It is capable of carrying 11 to 13 tons.

Three by Six.—Three strands with six manila-covered wires in each. The size of wire is slightly larger than the three by three, but the number of yarns and the laying up is exactly the same.

It is capable of carrying from 18 to 20 tons.

Seizing Wire.—Is made for all seizing, serving and binding purposes. It is galvanised and is 7-ply consisting of six wires around a central one which forms the heart.

MANIPULATING WIRE.

Opening a Coil.—A coil of wire must not be opened up in the same manner as a coil of rope or a multitude of "kinks" will be the result. Instead, it should be unrolled in the opposite way to which it was made up.

Small coils can be rolled along the deck in the same manner as a hose is unrolled, but larger ones require a turntable. No special turntable is kept for the purpose, so one has always to be improvised.

This is best done with two substantial pieces of wood lashed together to form a cross. Two bridles are attached by making an end fast on each leg of the cross, about midway between the centre and the ends, and the bights must be long enough to reach through the centre of the coil when it is laid on the wooden cross.

When the bridles are hooked to the cargo runner the wire is lifted clear of the deck, and it will revolve freely if the cargo hook is fitted with a swivel. If not, the wire must be landed on the deck at suitable intervals to take the turns out of the runner.

Measuring.—When sufficient has been taken from the coil, the exact amount can be measured off with a tape measure, or the required length may be chalked on the deck when more than one length is to be cut.

Don't forget to allow extra for the amount taken up by the splice. This will vary with the size of the wire, but as 24mm is about the largest that you are likely to work with, somewhere about 18 inches or 500mm extra for one eye splice would be necessary. Don't be niggardly about leaving yourself enough end though; a few inches extra will not be amiss and it is always best to be on the safe side.

Cutting.—Before cutting, a rope yarn whipping must be put on either side of the place to be cut, otherwise the strands will fly apart and spoil the wire for some considerable distance.

A few sharp blows with the edge of the hammer in the space between the whippings (about 1 inch) will flatten the surface, so that the cold set or chisel will cut more evenly and firmly. Cutting must be done on a good, solid foundation, such as the mooring bitts, or other heavy fittings.

SPLICES.

Notes on Wire Splicing.—Very little wire splicing is done on board the average vessel. To put an eye or perhaps a short splice in a mooring wire is about all a seaman is ever called upon to do.

All such work as making and setting up new standing rigging is done in port by skilled riggers, and even running gear is supplied ready for use.

(It is interesting to note that instead of a short splice in a broken mooring wire, some seamen put in a cut splice. This is thought to be easier and stronger, but does not look very neat.)

Wire splicing, like all other skilled work, is mostly given to the older "hands" to do, with the result that young seamen get few chances to practise marline-spike work. This is especially so with long splices, as they take a certain amount of skill to finish off properly and neatly.

As this form of splicing is seldom economical, any wire requiring to be done in this manner is usually condemned for the sake of safety and a new one is rove in its place.

Most seamen appear to content themselves with learning one or two eye splices, but more than that is desirable to meet all occasions likely to arise.

Many splices are in general use, but it is not necessary to know them all. About four will cover all ordinary purposes. For instance, a mate or bo'sun may insist on either "naval style", a locking splice finished off either right- or left-handed, or he may want the type stipulated in the Docks Regulations. Lastly, there is another splice which is seldom done on board, but which is always likely to be asked for. That is the one which is seen on the wire guards attached to the forks of the lamp brackets of the masthead lights.

The guides are spliced to the forks in such a manner that the usual thick splice will not hinder the passage of the cage while hoisting. It is known as a reduced splice.

In the course of time some eye splices have acquired a name, but generally speaking, they are usually designated according to whether or not they are locking splices or whether left- or right-handed.

In naval ships all splices must be left-handed, while in the Merchant Service any type is used and no fixed rule applies, except for the stipulation in the Docks Regulations.

Deck boys and ordinary seamen are sometimes sent to assist an older hand when splicing, but as these opportunities are few and far between, it is up to a boy to make the most of them when he gets the chance.

Eye Splices.—Two distinct types of eye splices are in general use, namely—left-handed, which is also known as "over and under", and right-handed.

Left-handed Splicing.—Tuck for tuck, left-handed splicing is considered to be much stronger than the right-handed style. The strands are practically plaited, binding the whole splice together, and the "bite" or frictional adhesion is much greater.

Right-handed Splicing.—This is the easiest and quickest method of splicing. Each end is simply wrapped round one strand about four or five times. Right-handed splices are all right for ordinary purposes, but are not much use if the wire is liable to spin and unlay.

Crane ropes and cargo runners sometimes spin a good deal, and for this reason, the Docks Regulations provide that a locking splice be used for such ropes and runners.

Locking Splices.—To form a "lock" will give added strength to either a left-handed or right-handed splice. This is done by tucking two ends under one strand in opposite directions, so that they cross. The lock is only made in the first round of tucks. Some types of eye splices have more than one lock.

Notes on Making Eye Splices.—The whole success of an eye splice depends on the first tuck actually made. If the first end is chosen and tucked correctly, the others will follow neatly in place; but if not, there is sure to be a long "jaw" through one or two strands leading farther than they ought to. The exact place to start can only be learnt by actual practice.

The first round of tucks is the most difficult; after that the rest is easy, being merely repetition work.

Length of Ends.—Good long ends for tucking are always advisable—short ones do not afford a good hand-hold when pulling tight, and they are more difficult to bend.

The proper length for the ends will depend upon the number of tucks to be made, but, as a rule, about 600mm is ample for the size of wire likely to be spliced on board the average vessel.

Whipping and Serving.—Measure off the length for the ends and pass a good firm whipping about 70mm long, working away from the end of the wire. This is done so as not to shorten the distance already measured off, and also to prevent the whipping slackening up when it takes the pressure of the unlaid ends on the first few turns. The first two or three turns are more secure than the last few.

Rope-yarn whippings are mostly used, but in better class work the wire is served over with marline for about the same distance 70mm.

Turning in a Thimble.—If a thimble is to be spliced into the eye, the serving is carried along the full length to be occupied by the thimble.

This distance may be measured by laying the grooved part of the thimble on the wire and turning it over once.

When ready to proceed, make the round end of the thimble fast exactly midway along the serving, bend both parts of wire together in the form of a loop, and squeeze into shape between the jaws of a vice, or with a rigging screw. Seize both parts firmly to the thimble.

Unlaying the Ends.—Place a light temporary seizing about 8 inches from the end of the wire before cutting the whipping at the extreme end. If this is not done the ends will fly apart and each one will open out and unlay. Whip each end separately with sail twine, working away from the ends, then cut the temporary seizing.

Cutting the Heart.—Divide the ends, force them apart, and cut out as much of the rope heart as possible. For the sake of neatness, the hearts are sometimes taken out of the strands also, but this is not a general practice.

Hints when Tucking.—Care should be exercised to avoid distorting the "lay" or natural set of either the standing part or strands when splicing. This may easily happen through unnecessary twisting or pulling when tucking ends, and is most likely to arise through trying to tuck an end too near the place where it emerges from the last tuck. To avoid this, give yourself more room by running the spike down the wire a few inches with a spiral movement. Follow up with the end and pass it through.

Next, work the spike back again to its former position, at the same time pulling the strand completely through, and haul tight.

Never hammer the strands back as far as they will go—except in the case of the first round of tucks. They should be allowed to lie in a long, natural "lay", without in any way being loose.

Starting the Tucks.—If the wire is a large one, or if more than one strand is to be picked up, a few tucks should be made before hanging the wire up in a convenient position for working.

It is easier to drive the spike through if the wire is lying on a solid foundation. Some seamen make the first round of tucks, hold all the ends together, then beat the standing part with a spike to tighten up all parts.

After the first few tucks, the usual practice is to hang the wire up in a horizontal position for ease in working the remainder of the tucks.

The beginner who keeps forgetting the number of tucks already put in should make a mark on No. 1 end—either with twine or a blow with a sharp instrument. This will save time wasted in continually tracing up and counting tucks.

Running the Ends Up.—After the first round, some men complete all the tucks with one end before passing to the next. This is a quick system, but there is a danger that the last strand will tighten up and become buried, but if the heart is not cut out it can be run up as well and the last strand will not be buried.

Number of Tucks.—No hard-and-fast rule can be laid down concerning the number of tucks to be put in a wire. Everything depends on the weight or strain it will be required to bear, or whether or not there is any "spin". Due regard must also be paid to the difference between right-handed and left-handed splices. In this connection it is considered that three left-handed tucks are the equivalent to five right-handed ones. Whether or not there will be a locking tuck is another factor to consider, and last, but not least, there is the wishes of the boatswain, or the chief officer.

Generally speaking, there is usually about $5\frac{1}{2}$ tucks (right-handed) in a cargo runner or a mooring wire, $4\frac{1}{2}$ in a guy pendant, and 3 or 4 in davit guys.

Half Tucks or Tapering.—This spaces the ends out more evenly, tapers the splice down, and gives it a much neater look.

To do this, tuck every second end in the ordinary way. That is to say, tuck Nos. 1, 3 and 5 ends. Alternatively, Nos. 2, 4 and 6 ends may be used instead.

Beating Out a Splice.—Before cutting the ends off, lay the completed splice on a solid foundation (such as the bitts), and beat it out with a heavy hammer. This will put it into shape, and any slack on the ends will be worked out. Beat from the neck of the splice to the ends, keeping the wire turning all the time.

Types of Eye Splices.—In the following splices it is assumed that the wire with the eye formed is laid in a horizontal position, end on to the person splicing, and that the unlaid strands are lying uppermost.

When the unlaid ends are laid out evenly in natural order they are numbered from right to left, and the strands of the standing part are numbered likewise, beginning with the first tuck, unless otherwise stated. As an aid to simplicity, a distinction has been made in type between the numbered ends and strands.

Locking Splice (Right-handed).—Pick up Nos. 1 and 2 strands adjacent to the first end (extreme right) and tuck it through towards right. Drop No. 1 strand and tuck the second end through No. 2 strand (towards right). This gives us two ends entering at the same place but emerging between different strands. Next, pick up No. 3 strand and tuck the fourth end towards right; then, without withdrawing the spike, tuck the third end towards left. This makes the cross or lock.

Pick up Nos. 4 and 5 strands, tuck the sixth end through both towards left, drop No. 5 and tuck the fifth end through No. 4 strand, also towards left.

Once again we have two ends entering between the same strands, but emerging between different ones.

Finish off right-handed.

This system of splicing usually leaves a large space between Nos. 3 and 4 strands. To counteract this, some men tuck the fifth end through No. 4 strand and the sixth end under No. 5 strand, both towards right.

Locking Splice (Left-handed).—Take the first end (extreme right) and tuck through nearest strand (towards right). Next, lift No. 2 strand (counting towards left), and tuck second end towards left. With the spike still under No. 2 strand take the sixth end (extreme left) underneath everything and tuck towards right.

Tuck the third end through No. 3 strand towards left, then pick up Nos. 4 and 5 strands and tuck the fifth end through both (towards left). Next, drop No. 5 strand off spike and tuck the fourth end through No. 4 strand (also towards left). This finishes the first round of tucks.

To continue, each end is taken over one and under one as in rope splicing. In other words, it is finished off left-handed.

Naval Splice (Left-handed).—Start with No. 1 end on extreme right and tuck through nearest strand towards left. Continue in the same manner with Nos. 2, 3, 4 and 5 ends, then tuck No. 6 through two strands, also towards left. This finishes the first round of tucks.

The second round is made by tucking each end over one and under one as in ordinary rope splicing.

3, 2 and 1 Style.—This is the most popular style of splicing in the Merchant Service, probably because it is one of the easiest to tuck and to remember, but it is not universal and hence the need to know more than one way.

To this method there is a number of variations, only slightly different from the original principle, and the same original principle is used on wires containing more than six strands.

Everything considered, it is a style well worth learning.

Without injuring the heart, insert spike down through centre of the wire from top centre, so as to have three strands on each side of spike and the heart on the left.

Tuck No. 1 end (extreme right) through the centre. This leaves number one, two and three strands on right side of spike. Next, drop number one strand (nearest point of spike) and tuck No. 2 end through numbers two and three strands. Drop another strand (number two) and tuck No. 3 end through number three strand.

We now have three ends entering through the centre of the wire but emerging between different strands.

After No. 3 is tucked and before starting to tuck No. 4, run the loose heart up around the heart in the standing part.

No. 4 end is tucked under number four strand towards right, No. 5 end under number five strand (towards right), and No. 6 end under number six strand, also towards right.

Finish off remainder of tucks right-handed.

Reduced Eye Splice.—Leave enough end for an eye splice with about four tucks, then pass a good whipping around the wire.

Unlay to this point and halve all the strands and heart as well.

Lay up one-half of strands to form a reduced wire, cut remaining half off and proceed with eye splice in the ordinary way.

The best plan when laying up the halved strands is to get a small, flat piece of wood about 4 inches square and bore six holes in a circle around a seventh one in the centre.

Poke the ends and heart through these holes, twist the wood around and the wire will lay up much tighter than by any other method.

The finished splice should be the same thickness as the wire.

Sometimes this splice is seen on the end of cargo runners, so that they may be attached more easily to a winch barrel, but it is more often used for attaching wire guides to the forks of lamp brackets, in order that the cage will run up and down the guides without jamming on a thick splice.

Short Splice.—Place a firm, narrow seizing about 18 inches from the end of each part to be spliced, unlay to that point and cut both hearts out.

Should the wire contain them, the individual heart may be extracted from each of the twelve strands before they are whipped with sail-twine. Although not absolutely necessary, this will help to reduce the bulk of the finished splice.

Interlock the strands as in rope splicing. This is done by alternating each end with one from the opposite side. Next, pass a strong seizing of sail-twine around all parts at the point where the opposing strands meet and cross.

With the "job" held thus in position, the first round of tucks can now be made on either side. In the meantime, the two seizings which keep the wire from unlaying must be left in place. Consequently, the first round of tucks must jump these seizings.

The same method as in rope splicing is used, over one and under one, the only difference being that a total of twelve ends are involved instead of six. When sufficient tucks have been made the first two seizings must be cut and picked away.

About three or four tucks each way should be enough for all ordinary purposes, and if desired, the splice may be tapered.

Next, cut the central seizing at the junction and beat into shape; then cut off surplus ends and finish off with another beating.

Long Splice.—Experience in making a long splice in wire rope is difficult to obtain on board any class of ship. Only the most skilful hands are allowed to attempt it, for the amateur may spoil the job and waste a good deal of wire. So much end is involved that it would be impossible to get sufficient old wire to practise with, as in the case of eye splices, but the learner can, if he is anxious to learn, practise finishing off the ends on any short piece of old stuff available, for this is the most difficult part of the job to learn.

Galvanised wire is never used for running-gear, as the galvanising would soon wear off through contact with the sheaves. For this reason it is used mostly as standing rigging, and even if such rigging does carry away, it is not likely to be spliced, so that flexible steel wire is almost sure to be the material used when a long splice has to be done at sea.

In planning the amount of end to unlay, due regard must be had to the size of the wire, and the distance apart you wish to have the meeting-places of the strands, when both ends are brought together and the strands interlaid. Opinions vary as to the actual length of end required for a safe splice, so the figures given below are meant to cover the range in general use, but care should be taken to allow plenty. To be niggardly may spoil the whole job.

Unlay from 10 metres to 12 metres for a wire 28mm in diameter.

Unlay from 4 metres to 5 metres for a wire 12mm in diameter.

We will take, for example, a 12mm wire. Measure off 4 metres on each end and pass a good whipping at these points. These we will call the No. 1 seizings.

Some men prefer to open out the strands in pairs, but the easiest method is to unlay three alternate strands singly, which leaves us with three strands still laid around the heart. Cut the heart as well as the remaining strands about 150mm from the No. 1 seizing, then open out the short ends to that point.

We now have three strands 4 metres long, with a short 150mm strand lying in between each of the long ones.

When both wires have been completed in this manner, interlock the long ends in such a manner that a long and a short strand come together. With assistants pulling in opposite directions to hold the wires together, cut both the No. 1 seizings. See that the two short ends of heart are hanging together and clear of the join, then pull strongly in both directions to make sure the wires are jammed together as tightly as possible. If this is not done a very unsightly join will be the result.

Take the three ends belonging to one side and tie them securely to the standing part, which leaves us free to interlay the three short and three long ends on the opposite side.

Unlay one of these short ends, and at the same time lay in the long end which lies nearest to it. Continue this until about 1 metre of the long end remains, then cross both parts to lock them. Measure off 1 metre on the unlaid strand, whip and cut off superfluous end. The point where these two strands meet and cross will be about 3 metres from the join.

Unlay the next short end and lay in the long strand for a distance of 2 metres, then cross them. Leave 1 metre of end on both parts, then whip and cut off the remainder.

Do the same with the third pair for a distance of 600mm from the join, then finish off in the same way.

This finishes one half of the laying up, so we turn to the opposite part of the splice and proceed as before.

When this is finished there should be six equidistant points, at each of which two strands meet and cross. The meeting-places will be 1·3 metres apart, and all the ends which are now ready for tucking will be 1 metre long.

We have now arrived at the hardest part of the job; the part which really requires experience before any degree of proficiency is obtained.

The problem now is to remove the heart and bury the remaining ends, in order that they will take its place and act as a central core to keep the outer strands in place.

Burying the Ends.—As the outer wires would not grip on the smooth surface of a buried strand, and in order to bring them up to the same thickness as the heart itself, all the ends to be buried are either parcelled with bagging or served over with thin yarns. Parcelling is best, but before it is put on the strands should be cut the exact length for butting together inside the wire. Each will require to be about 600mm but they must come together exactly or the outer wires will sink into the space left vacant. If too long the wire will develop a bulge. Fig. 38.

Start tucking in the centre. Pull one part of the heart out of the wire, to a point just past the first pair of ends, the nearest of which will be the first end to be buried. If the heart should be difficult to pull out it can be helped considerably by running a spike along and through the strands.

When ready to bury the first end, insert the spike from the opposite side of the wire, picking up one strand and the remaining end which lies nearest to it, in such a manner that the point of the spike covers the one to be tucked. As the spike is moved along in the lay of the rope the end should fall into the inside of the wire.

At this point we will have to decide whether the ends should be made to cross before disappearing inside the wire, or simply lie side by side instead of crossing.

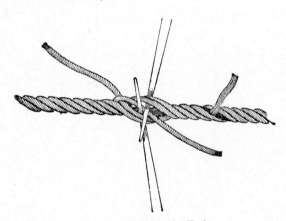

FIG. 38.—Burying the End.

This point will be determined by the formation of the strands themselves.

If the wires of a strand are twisted up in the opposite direction to the lay of the rope, then the ends should be made to simply disappear into the wire without crossing, forming what is known as a "flat joint".

On the other hand, if the wires are twisted up in the same direction as the lay of a rope, the ends should be crossed to form a round joint.

If a round joint is to be made, two spikes will be required to make a sharp, neat "cross".

Pick up two strands, from opposite directions, on each side of the lay where the end is to disappear, and let the strand which is about to be tucked lie between the ends of the two spikes.

By moving the spikes in opposite directions the strand will be crushed into the wire. When well clear of the joint, one of the spikes may be taken out and the remainder of the tucking done with the spike which covers the end being buried.

When all the ends have been buried, a little hammering and twisting will put the wire into shape and it will then be ready for using again.

Wire Grommets.—A strand extracted from a length of flexible steel wire is the best material with which to make grommets. A completed grommet should contain five strands, which is the most that can be laid up by hand; consequently, the length of the single strand should be at least five times the circumference of the proposed grommet. The "lay" of the wire will use up some of this length, and extra must also be allowed for finishing the ends off. Therefore, about five-and-a-half or six times the circumference would be a good measurement.

It is much better to have plenty of end than to find out later that it is too short.

Start the grommet with an overhand knot, taking care that the strand crosses in a natural "lay". When a few turns have been laid up, the circle can be tested for size, bearing in mind that parcelling or serving will reduce the inside circumference still more. Lay the strand up from left to right.

Finishing off the Ends.—Various methods may be employed for finishing off the ends, but the simplest way—if the grommet is to be served—is to whip and cut them, so that they will butt end to end. A more substantial method is to bury the ends in the same manner as for a long splice, in which case extra length will be required in the original length of the strand.

Another method is to halve each end, cross the parts, then dispose of them by tucking over one and under one.

Stropping Blocks (with Rope).—Only small blocks are likely to require stropping, and they are too small for a strop joined with a short splice, so it will be necessary to make a grommet.

When measuring the length of strand required for the grommet, the same rule may be used as for wire; that is, once round the block and thimble for each strand of the completed grommet. A three-stranded rope will give us three times round everything, but the lay of the rope is shorter than that of a wire, so it is best to allow an additional 300mm, in case it is required.

Care should be taken to allow for serving (if it is to be put on), as this will reduce the inside circumference quite a lot.

Stropping Blocks (with wire).—Large-sized blocks may be fitted with a strop joined by means of a short splice, but the only blocks likely to be stropped are so small that it would be necessary to make a grommet.

A spliced strop is more economical, but a grommet made with a single strand has the advantage of looking much neater. The length of wire required can be measured with a piece of small stuff of similar thickness to the strand which will be used, and this should be unlaid from a length of six-stranded flexible steel wire. The general rule for measurement is once round everything for each strand of the completed grommet.

Five times is the most that it will be possible to "lay up" the strand. Therefore, the measurement will run as follows—five times round the block and five times round the thimble, plus about 300mm extra to allow for the serving.

When starting the first few turns of the grommet great care should be taken to see that it will be large enough, bearing in mind that the serving, when put on, will reduce the inside circumference of the strop a good deal.

Fixing the Strop.—Place the strop round the block in such a manner that any slight lumpiness will be between the block and the thimble where the seizing is to be applied. If a rigging screw is available it should be used for heaving the parts close in to the thimble, but failing that a Spanish Windlass will be found handy. To simplify matters the block should be hung up with a lashing through the thimble and another through the swallow of the block. This will keep the strop in position while the seizing is being passed. Don't forget to place a piece of canvas around both parts to prevent the seizing separating or cutting into the serving.

Worming.—A method of filling in the grooves between the strands of a rope with marline or other small stuff. It is seldom seen nowadays, but was once much used on fire-bucket handles and manropes on accommodation ladders.

Parcelling.—This serves the double purpose of keeping the weather out of a wire, and helps to level off any unevenness near the region of a splice. Old bagging or burlap makes good parcelling; it should be cut in long strips about three inches wide and rolled up ready for use in the same manner as a bandage. A thick coating of tallow is put on the wire, and the parcelling is laid on in the same manner as a soldier's puttee. To hold it in place a length of sail twine is hitched along the full length of the burlap. If another coating of tallow is now applied over the burlap, it will both help to assist the movement of the serving-board and keep the weather out as well.

Don't forget to put the parcelling on with the "lay" of the rope.

Parcelling Stays.—The idea of parcelling is to keep water out of the wire. Therefore, when working on a stay it should be put on with the "lay", starting at the splice on each side and finishing off at the middle of the eye.

This overlapping keeps water out, and for the same reason, swifters, shrouds and stays are parcelled upwards.

Serving.—This is the finishing touch which makes all the difference between a neat job and an eyesore. After an eye has been spliced in a wire, the part where the ends have been cut short is parcelled and served. Not only does this keep the weather out, but it also protects the hands from the short, sharp ends of the cut strands.

Two- or three-stranded marline is invariably used for this job, and to bind it on the wire sufficiently tight a serving-board is used. These are of two kinds—round and flat. The latter is the most popular type; it is grooved out on one side to take the shape of the wire.

Before starting, remember that serving must be put on against the "lay" of the rope, therefore it would be started at the end of the splice and worked towards the eye itself.

The wire to be served should be hung up at a convenient height, and the person serving should stand with the eye of the wire on his right-hand side. The person who is assisting by passing the ball of marline around the wire should stand on the opposite side. To begin, unlay the end of marline about half-an-inch or so, and put a few turns on by hand. The end will be under the first few turns, and being unlaid will not leave a bulge. The serving-board should work away from the person serving, therefore, the grooved part is laid against the wire with the rounded part towards the body.

The marline is laid up the rounded side from the bottom, goes round the handle, back down the same side and is then led up to the handle again. When working, the marline runs through the fingers to regulate the strain put on the turns. The last four turns are made large enough to allow the ball of marline to pass through, so that when pulled tight again an ordinary whipping is formed. An overhand knot is sometimes worked down flush with the serving before cutting off.

Joining Serving.—When the marline is nearly finished—say about a metre from the end—lay end of new ball along the wire and serve over it for 100mm. When old stuff is finally finished, pass remainder under new marline and carry on serving over the old end.

Wire Serving-Board.—An ordinary type of serving-board is of little use when seizing wire is employed. In fact, the practice is so uncommon that no special board is kept for the purpose and one usually has to be improvised.

Take a piece of hard wood about $40 \times 5 \times 3$cm and bore three holes, each 6mm in diameter, in the following order: the first about 30mm from the end, and the second about 50mm. The third one should be spaced slightly more than the diameter of the article to be served.

CANVAS.

Canvas.—The word canvas is thought to be derived from *cannabis*, a Latin word for hemp, which would seem to indicate that in the early days of its history it was made from that material. The Egyptians and Phoenicians are also believed to have used canvas, and were probably the originators of this type of cloth.

Construction.—At the present time canvas is made from hemp, flax, cotton, tow and jute, or a mixture of these materials. British canvas is made from flax, but in America cotton is mostly used, and this gives rise to the term "American cotton canvas". Best quality British canvas is made with "all long flax". Bagging, tarpaulin and canvas, in the order named, form an ascending series of cloths as far as fineness is concerned.

The threads running lengthwise along the canvas are known as the "warp", while those running across the cloth are called the "weft". On account of the rough usage it gets, and the strong pressures it must stand, all flax canvas is made with a double warp.

The finished edge of the canvas is called the selvedge, and the coloured thread woven near it is used as a guide for the seam when sewing lengths of canvas together, and is known as the selvedge stripe.

On the outside of each bolt of canvas, the length to the nearest quarter of a yard will be found stencilled thereon. The width is 600mm.

Waterproofing.—Most grades are proofed to prevent water percolating through the woven material, and two methods of doing this are in general use. In the first, the fibres are chemically impregnated, and in the second a wax proofing is superimposed on the material itself.

The former is probably the best, as the action of the sun and continual folding will destroy the efficiency of the proofing in the case of the wax surface type.

Grades and Uses.—Hundreds of different types and sizes of canvas are to be found on the market, but that in general use in the Merchant Service is known as "Merchant Navy Canvas", and is supplied in seven grades. Each grade is known by a number from 0 to 6, and each is used for a particular purpose, as under:

No. 0 is a very stout canvas used principally for making canvas
 ash shoots.
No. 1 is used for awnings.
No. 2 is used for awnings, boat covers, winch covers, etc.
No. 3 is used for awnings, boat covers, winch covers, etc.
No. 4 is used for awnings, ventilator covers, screens, etc.
No. 5, soft, cheap line used for anything, principally patching.
No. 6, soft, cheap line used for anything, principally patching.
All above materials are 600mm wide.

Sail Needles.—Sizes run from 6 to 16.
6 to 13 are roping needles.
14 to $14\frac{1}{2}$ are seaming needles.
16 are duck needles.

Tarpaulin Canvas.—As a general rule tarpaulin canvas is rougher
and coarser than ordinary canvas. It is well proofed to withstand
water and is made from second-grade flax. Although not of such
good quality as ordinary canvas, yet some of the finest grades of
tarpaulin are of better texture than the lower grades of canvas. It
is manufactured in both green and white, in widths of 750mm
and 900mm.

Yacht Canvas.—Although the terms sailcloth and canvas are
synonymous, it is really only yacht canvas or material for lifeboat
sails which is spoken of as sailcloth. This is supplied in 300mm,
400mm and 500mm approximate widths, but "terylene" is now
being used increasingly for sails for yachts.

Cotton Canvas.—Like other types of canvas, the cotton variety
can be obtained in many grades and sizes. It is manufactured and
used principally in America. As a general rule, it is harder to sew
than flax canvas.

Duck.—This is a very light type of cotton canvas. It is manufactured in hundreds of different grades and sizes, and is only used
in best class of work, when a more pleasing effect on the eye is
required.

Canvas Work.—Apart from neat stitching, the most important
point to watch in canvas work is measurement. By allowing the
correct amount for shrinkage and stretching, even an amateur
should turn out a creditable job. Sailmakers allow one inch to the
foot for "stretch" when making awnings, but for boat covers or
tarpaulins the same amount would be allowed for shrinkage.
 For smaller articles, less than one inch to the foot is often
advisable, but this is purely a matter of judgment, and depends
on the article being made.
 Another point to watch is the fact that the bottom cloth always
creeps in or shortens when two cloths are being sewed together. If

both were cut the same length, this shortening would probably spoil the job; therefore, an allowance must be made to counteract this.

The overlapping of flat seams must also be taken into consideration, for there would be considerable loss in this respect on a big job such as an awning, and even with small articles the loss would be important.

Strike Up Marks.—Pencil marks across a seam before it is sewn. If these move apart "creep" is obvious.

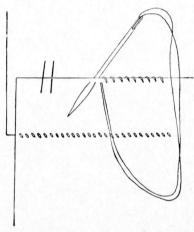

FIG. 39.—Flat Seam showing Strike Up Marks.

Sewing Canvas.—Two different types of seams—round and flat—are used when sewing canvas. The type of work, as well as the purpose the material is intended for, will determine which one should be used.

Flat Seams.—Overlap two strips (cloths) of canvas as far as the coloured marking thread, and sew on both sides. This makes a double row of stitches and a very strong seam.

When sewing any article which will be exposed to the weather, the seams must be overlapped so that they will not catch any wind or rain, in the same manner as slates on a roof.

Begin sewing on the right and work to the left. Fig. 39.

Round Seams.—This is the quickest method of joining two pieces of canvas. The stiches have to be made on what will eventually be the inside of the article; therefore, the parts are reversed before sewing together.

Turn down about half-an-inch of each edge, and rub it down with the handle of a knife (or anything hard), to form a sharp, distinct fold.

Place the two parts of canvas together, side by side, vertically, with the short ends or folds turned away from one another.

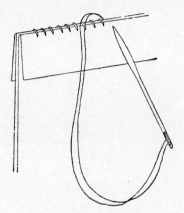

FIG. 40.—Round Seam.

Sew from left to right. When finished, open out the canvas and rub the seam down to flatten it out. Fig. 40.

Tabling.—Nearly all articles are finished off with a "turn in" or fold sewn along the edges. This is known as a "tabling". Fig. 41.

The double thickness of canvas gives a good strong edge, especially if eyelets are to be stamped in, or if roping is to be sewn on.

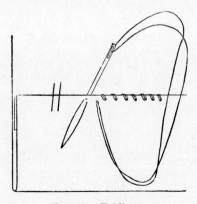

FIG. 41.—Tabling.

Stitches.—Very neat sewing requires about five or six stitches to the inch, but four is the commonest number in ordinary work.

Anything less than that would give the appearance of "Homeward Bound Stitches" or "Dog's Teeth" (e.g. big, careless, hurried stitches).

Beeswax.—Before using twine for sewing it is usually rubbed down with beeswax or soap. This smooths down the fibrous surface of the twine and helps to make a more watertight seam.

Unsmoothed or unwaxed twine will quickly wear through and break.

Alkali in soap is injurious to twine if used too freely. In fact, it should not be used unless absolutely necessary.

Starting the Stitches.—Among "old-timers" it is considered to be bad form to start sewing with a knot in the end of the twine.

The proper way, they say, is to leave a short end (about 40 mm), tuck this in along the seam, then sew round it.

Joining Thread.—Various ways of joining thread after re-threading the needle are in general use, but the following method is as good as any:

Leave about $1\frac{1}{2}$ inch of end to the canvas when cutting the needle adrift, then unpick the last half of the stitch. Begin sewing again by sticking the needle through the hole left by this unpicked stitch. If a short end is left on the new length of twine, we will now have two short ends coming together inside the seam. Twist these ends together, lay them just inside the seam and sew around them.

Another method is to proceed as before, knot the fresh twine, and continue sewing around the end from which the needle has been cut.

Eyelets.—Those used in canvas are of metal and they are hammered in with a special punch.

Working an Eyelet Hole.—Make a small grommet and place it around the hole. Commence sewing at the point farthest from the person by sticking the needle down through the canvas and passing it up through the hole. Alternate stitches are carried well out from the hole and the grommet must be completely covered.

Repair Work.—A cut or tear in the canvas which is not worth patching is drawn together by "Herringboning" or "Cross Stitching".

Cross Stitching.—At the left-hand end of the tear, stick the needle (with a knot in end of twine) up through far side of canvas, then stick through and down on the near side, and to the right, giving the twine a wide angle, then across to the far side and up through, also at a wide angle.

Lead the twine back across the first stitch at an even angle, to form a neat X, then push the needle back at the same angle until it emerges again at the same place as the twine. Next, take another stitch to the right and carry on as before.

By reversing the procedure it is possible to work from right to left.

Herring-boning.—Starting from the near side of the tear, and at the right-hand end, make the first stitch up through the opposite side, at an angle to the left, then lead back to near side, pushing the needle down through and bringing it up on the right side of the crossing stitch. Lead across to far side at an angle again, and so on, for the required distance. Fig. 45.

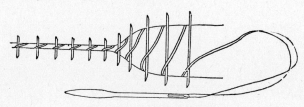

Fig. 42.—Herring-boning.

When covering rails, this system counteracts the tendency of an ordinary seam to spiral around the rail, through continually drawing to one side.

Darning.—This is done in the same way as darning socks, but can only be used when the hole or tear is a very small one.

Roping.—Hold the roping on the far side of the canvas. Stick the needle through the canvas and through a strand, lead the twine

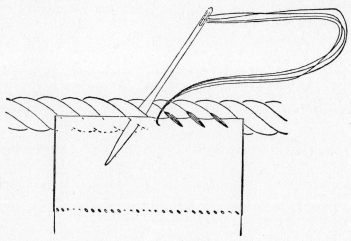

Fig. 43.—Roping.

down between the lays of the rope, then through the canvas again and through the next strand, working from left to right. Carry on in this manner, and to finish off an end of twine, stick the needle back once in the opposite direction from which it came, thus forming a short nip. Fig. 43.

Roping is invariably sewed to a tabling—very seldom to a selvedge edge.

Stowing Away Canvas.—On no account must canvas be stowed away while wet, or even damp, or it will quickly rot.

In addition to the markings which should be on the outside, if the article is made up properly, there should also be tags with the proper markings tied on as well.

BOSUN'S CHAIR.

Lowering Hitch.—A bosun's chair and gantline is used in all work done aloft, and the gantline is always attached to the chair by means of a double sheet bend, with the end stopped securely to the standing part of the gantlines. When painting topmasts or "Riding Stays", another man always attends the gantline and lowers the one who is painting, but when working on the lowermast, a sailor is expected to lower himself by means of a lowering hitch. This takes the form of a reef-knot on the chair itself. It is formed as follows (Fig. 44):—

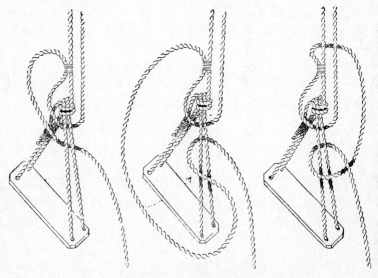

Fig. 44.—Lowering Hitch.

When hoisted high enough, a wracking is passed around both parts of the gantline, and this holds the weight while the hitch is being made. When ready, a long bight of the hauling part is pulled through the strop of the chair towards the person, passed over the head, and allowed to drop behind to the feet, which are passed behind it. When the sides of this long bight have been brought to the front of the person, the slack on the hauling part is pulled tight, and this forms a reef-knot on the strop of the chair itself. When ready to lower away, the wracking is cast off, some slack is pulled up, and the parts are made to render round, which they will do easily enough through the weight of the man in the chair.

Fig. 45—Alternative Method for Lowering Hitch.

Instead of using a wracking, most seamen simply pull a little slack through the strop and lead it up to the standing part, both parts of which they hold in one hand while forming the hitch with the other. This is a quick method, but it should not be attempted until the forming of the hitch is thoroughly understood, and has been practised a few times.

Another good method of lowering oneself, when sitting in a bosun's chair, instead of using the usual lowering hitch, is to employ a good size hook, either one with an "eye" or an ordinary meat hook, or even a strong pot hook. Fig. 45.

The idea is to fasten the hook firmly to the standing part of the gantline so that the hauling part can be passed under the strop, up and over the hook, under the strop once more, then up and over the hook from the opposite side.

The standing part of the gantline may be rove through the "eye" of a hook, if so fitted, and made fast with a bowline, or it may be simply seized in place with a piece of marline passed through the lay as well as around the gantline. With a meat hook or pot hook the point should be run through the lay and seized in that position. See also Fig. 53.

Lowering a Telescopic Topmast.—This operation is often referred to as "striking the topmast". Apart from an accident, perhaps the only time it is ever necessary to lower one down is when bound up the Manchester Ship Canal. As it is so seldom done, everything is usually covered over with paint and rust, and it may be necessary to chip some away if everything is to work smoothly.

All gear, such as the fore topmast stay, backstays and jumper stay should be slackened up, and the wireless aerial lowered down altogether. It will also be necessary to send down the masthead lamp gear if it is fixed to the topmast. A heel-rope must next be rove over the sheave on the side of the lowermast, through a sheave in the heel of the topmast, and then shackled to a lug on the foreside of the lowermast. A small length of wire is always kept rove permanently, so that it is only necessary to marry the heel rope to it, pull through and shackle to the lug. When the mast coat has been removed, and the wedges between the topmast and lowermast are taken out, the weight is taken on the heel rope, and the fid which supports the weight of the topmast is then withdrawn. The topmast is then lowered right down inside the lowermast.

Making Strops.—Any length of rope with the ends joined together will form a strop. A temporary one may be made by knotting the ends, but those intended for some special purpose are always spliced. A short splice is used, and the ends are tucked two or even three times each way, according to the strain it is intended to take. Before splicing, the rope should be stretched along the deck and all turns taken out, otherwise the completed strop will contain a few permanent "kinks". The length of a strop is the length when spliced. A sailor sent to make a two-fathom strop would require a little over four fathoms, which would allow extra for the splice. A good, strong strop can be made in a hurry by joining the ends together with a shroud knot.

Marrying Ropes.—A sailor sent to reeve off new running gear would not unreeve the old rope first, and then perhaps have to climb aloft to reeve the new gear. Instead, he would "marry" the end of the new stuff to the end of the old, in such a manner that it would pass through a sheave. As he unreeves the old gear he would automatically reeve the new. "Marrying" must be done neatly or it is liable to jam in the sheave. Before marrying, each end must be firmly whipped, and then a long rope yarn is passed through a strand on one of the ends. Pull the yarn through until both ends are equal. With the two rope ends butted together, pass the two yarns across and underneath a strand on the other rope. Repeat this operation three or four times, then separate the yarns and bind them round all parts, crossing them at each turn, and working towards the opposite rope where they are made fast together.

MOORINGS.

Mooring Ropes.—About three or four mooring ropes, and the same number of wires at each end of the vessel is the normal number usually carried, and for convenience they are always coiled down near the warping winch or mooring bitts, but never on a bare iron deck. Thick dunnage or properly constructed gratings are placed under them, and they are kept well clear of any exposed steam pipes. Fig. 46.

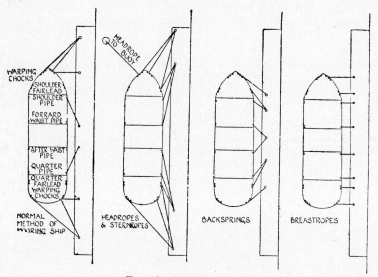

FIG. 46.—Mooring Ropes.

Kinks.—A multitude of kinks will form in a rope if it is continually led over the top of a winch barrel, but this can be avoided if the winch is occasionally reversed and the rope led underneath. Kinks in the after lines can be easily removed by towing over the side, but permission to do this is not always given. When heaving the bare end on board, heave the eye right over the barrel, if time permits, otherwise a few kinks will needlessly be left in the rope.

Surge.—This is an order frequently given to a man heaving on a rope. To do this, take a turn off the barrel, and let the hauling part hang slack in the hands. This allows the barrel to revolve while the rope remains stationary. Man-made fibre ropes may melt owing to the heat generated by friction and great care must be taken to avoid accidents.

Riding.—When turns on the barrel become crossed and lead over one another, the winch must be reversed to clear them.

Inside Turns.—Very often, through a bad lead to the barrel, an inside turn has to be taken to avoid riding. This brings the hauling part to the inside of the barrel and the standing part to the outside.

Putting Out a Bight.—After a vessel is hove alongside her berth a number of ropes and wires are put out to make her fast, and in addition to single ends, a bight of the same rope is often put out as well. To do this, heave the single end fairly tight, and take a few temporary turns on the bitts, pass the bight ashore, but make sure that the hauling part is in the same lead as the standing part, and that the third part is in a separate lead, otherwise one moving part would jam the other. When ready, throw off the temporary turns and heave all parts tight.

Rope Stoppers.—In order to transfer a mooring line from the winch to the bitts, it must be "stoppered off" to hold the weight, while the transfer is being made.

To attach the stopper, a single half hitch with the end backed two or three times is very often sufficient.

Chain Stoppers.—As a rope stopper cannot grip a wire properly, a chain stopper is always used for the purpose. To make fast, either two or three half hitches are formed, about 18 inches apart, and the tail is backed around the wire itself.

Note.—It is better to form a widely-spaced "cow hitch", otherwise, two half hitches are liable to slip together and jam.

"Come Up to the Stopper".—As soon as the stopper is made fast, this order is given as a signal to the man at the winch to let the rope go, but he must do this very gently, for a sudden jerk would probably part the stopper. In fact, it is a much better plan to reverse the winch, especially if a heavy strain is on the rope.

Messengers.—When a heavy wire or rope, which will not go round the winch barrel, is to be hove in, a smaller rope known as a messenger is made fast to the larger one, which is hove in bit by bit.

Chafing Gear.—Mooring lines leading across sharp surfaces are parcelled over with thick bagging or burlap to prevent chafe.

As the line is liable to stretch or slacken, parcelling should be carried well above and below the affected area, to allow for any alteration of position.

Singling Up.—Some time before leaving dock, all lines which can be done without are hove on board. Usually the moorings are reduced to a single headline and backspring forward and a single stern line and backspring aft so that no time is lost when it comes to sailing.

Lashing Mooring Ropes.—As a precautionary measure, mooring ropes are often lashed down owing to weather conditions, and for this purpose rope stoppers and heaving lines are mostly used. These are passed under as much of the rope, and round as many nearby stanchions and suchlike, as possible. In addition, the top eye of the rope is placed over the nearest bitts.

Stowing Away.—At sea, when not in use, all mooring ropes are stowed below decks, both for the purpose of preserving them and also to keep the decks clear. All stoppers, fenders, heaving lines, messengers, etc., are usually thrown on top, so that they will be handy when wanted again.

Making Up Gear.—Too much emphasis cannot be laid on the fact that gear of all kinds should be "made up small" (e.g. folded or coiled as the case might be).

Space in storerooms and lockers is always limited, and gear which is only used occasionally will take up less space if made up small and neat. Gear of this type (gantlines, stage-ropes, awnings, etc.) should always have labels attached to prevent unnecessary mistakes when searching for something that is required in a hurry.

Mousing.—To prevent a hook jumping out of an eyebolt aloft, a few turns of spun yarn are passed around from the point to the back of the hook, and made fast tightly. Fig. 47.

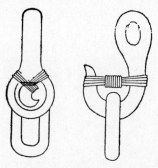

Fig. 47.—Mousing Hook and Shackle Pin.

If the mousing is to be permanent, a strand of seizing wire would be used instead of spun yarn. Much the same thing is done with shackles which are inclined to be slack. A few turns are taken through the hole in the pin and passed round the side of the shackle.

SOME SAILORISING JOBS.

Pilot Ladder.—Notice M898 sets the minimum standard for pilot ladders and their repair.

About 20mm hemp or manila is the usual size of rope used, and this should be well stretched for about twenty-four hours before being used. The two parts of rope on each side of the ladder are all one piece with the bight at the bottom; therefore, the full length of rope required will be double the length of the proposed ladder. 2 metres must also be added to this length for use as a lanyard to hang the ladder by.

Begin the ladder by measuring off the lanyard on each side rope, and mark the place with sail twine. The other end of the side rope should then be brought level with this mark, and if thimbles are to be fitted to the bottom, now is the time to seize them into place. When this is done, reeve all the steps on the side ropes, and they will now be ready for spacing and seizing.

For convenience in working, the ladder should be stretched out tight with a tackle attached to the lanyard ends, at a height above the deck of about 1 metre.

A seizing will be required on each side of all the steps, and once the proper length of seizing is found, a large number may be cut ready for use. If the completed ladder is to look uniform and neat, the same number of turns must be made in each seizing.

Two men working opposite one another is the best method at this stage, and each should be provided with a length of wood cut to the size allowed between the steps, so that they may be measured accurately.

When the last step is in place, the short ends of the side ropes are spliced into the longer ones, which are the lanyards, and these should be finished off with palm and needle whippings.

Rattling Down.—Most modern vessels have dispensed with ratlines altogether, but where they are still fitted, constant attention is required if accidents are to be prevented; therefore, old ones must be replaced from time to time as occasion demands.

To renew a ratline on the port rigging, take the end of a coil of 18- or 21-thread ratline stuff aloft after splicing a small eye in the end. Make a loose clove hitch on both the second and third shrouds from forward (keeping the second part of the hitch on the bottom), and leave enough end to stretch to the fourth. The eye on the end is now made fast to the after shroud with marline, and is kept in a horizontal position until the seizing is finished. Make the marline fast to the eye, pass it around the shroud to the opposite side of the eye, then backward and forward in like manner until a substantial seizing is formed. The end is finished off by hitching between the eye and shroud.

With the after-seizing finished, tighten up the ratline and cut off by the forward shroud, after allowing enough to splice another small eye. This end is also seized to the shroud in the same manner as the after one.

In the starboard rigging the ratline would be started on the forward shroud.

Reeving Relieving Tackles.—For small vessels relieving tackles may be used to take the strain off the quadrant and steering-engine

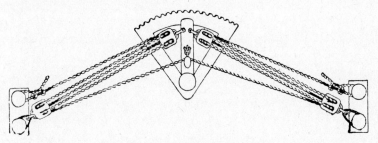

FIG. 48.—Reeving Relieving Tackle.

during bad weather. Heavy seas striking the rudder, combined with a certain amount of "play" on various parts, would soon cause a breakdown.

What may be described as two tackles are used, one on each side of the quadrant. The fall is rove through all the blocks, and when set up good and tight, permits the quadrant to move without excessive "play", the friction in the sheaves acting as a shock absorber. Fig. 48.

Reeving Runners Aloft.—While a derrick is up in position, it is often necessary to unreeve an old cargo runner and put in a new one in its place.

To do this properly, a heaving line is married to the end of the old one before it is unrove, then the line is married to the new one, so that it can be hove up and pulled through the block. A good force may be necessary to pull it through, so that the line must be married very firmly to the runner, and the best way to do this is as follows:—

Unlay about 100mm of the rope, form a wall, butt the ends and lay the strands along the wire, then serve them over tightly with a yarn.

D

Riding Stays.—Two methods of bending a bosun's chair to a gantline are in general use when riding stays. The first is to make the gantline fast with a double blackwall hitch, leaving a long end which is passed through the shackle on the stay and made fast on the standing part. The safest method is to put a good seizing on the strop to form a small "eye", bend the gantline to the strop below the seizing, and shackle the eye direct to the stay. Fig. 49.

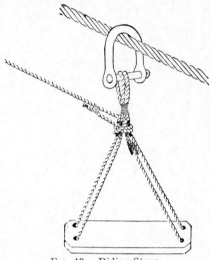

FIG. 49.—Riding Stays.

Before sending a chair aloft, make sure the bow of the shackle is riding on the stay, instead of the pin, which may revolve and come out.

Rigging Stages.—Practically any size line can be used as a stage rope, but 20mm is the handiest. It must be long enough to reach the waterline on the bight when painting the ship's side, and if the boot-topping is to be done, it will require to be still longer.

A large marline-spike hitch formed on the bight of a stage rope and slipped over the horns is the usual method by which ropes are attached to a stage, but so many young seamen experience difficulty

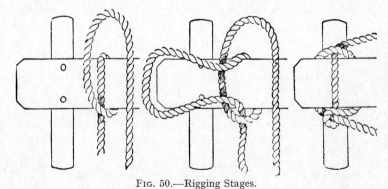

FIG. 50.—Rigging Stages.

with this method that an easier and quicker way of doing the same hitch is given here. Fig. 50.

Find the middle of the stage rope, then stand at the side of the stage facing the horn, with the end of the stage on the left.

Hold a part of the bight in each hand, then take a turn under the stage and inside the horn with the part held in the left hand.

Bring both parts together and cross them by taking the right part over the left, in the same way as for an overhand knot.

Drop the right-hand rope, then pick up the adjacent left-hand one. Lead it over the part just dropped and pass it over the end of the stage. Adjust and tighten by pulling on both ends.

Another easy method is to middle the rope as before and stand with the end of the stage towards the person. Lay the single part of bight across the stage, on the far side of the horn, cross both parts underneath, bring them up each side of the stage (on near side of horn) and cross again on top. Lift the first bight, dip over end of stage and haul tight. Fig. 51.

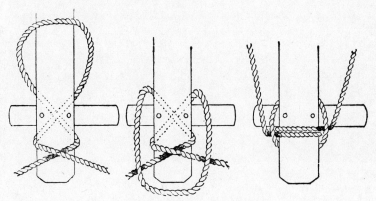

Fig. 51.—Rigging Stages, Second Method.

If the stage should be an extra long one, it is advisable to rig a centre line to prevent it sagging in the middle.

Instead of climbing on board each time the stage requires lowering, it is sometimes possible to pass the ropes around a stanchion or rail and lower from the stage itself. The best rig of all is two tail-blocks and single-part stage ropes, if these can be procured. Failing this, a lizard or a shackle and a lanyard will be found quite effective.

Rope-yarn Swabs.—These are usually made specially for wetting wooden decks with caustic (or other strong cleaning agent), when they are to be barberised. They take the place of ordinary mops, as these would soon be ruined by the action of the acid.

Take a light pole about 1500mm long and cut a good, deep notch around one end. This will be the shaft.

Rope-yarns about 800mm long are a good handy size to use, but they may be longer or shorter if desired.

Separate the yarns, make two bunches, and lay them along the pole against the notch, so that they overlap the end about 400mm. Pass three good seizings to hold them in place, turn remainder of ends down, and while holding them firmly in place, finish off with two more seizings.

Sail Twine—Preparing for Use.—A skein of twine is made up in the same way as wool when it first comes on board, but it requires to be put in handier shape before it is ready for use. Stretch the skein to its fullest extent. Cut and middle it, then make three bunches of the twine and plait these loosely together. Put a light seizing at the bottom to hold the ends, and one on the bight to hang it up by. Single threads are pulled out from the bight.

Sending Tail Block and Gantline Aloft.—The block is usually carried over the shoulder by means of a bowline formed on the tail. If the climb is not too high the gantline is rove in the block and a figure-of-eight knot made in the end.

Alternatively, both may be sent up on the signal halliards.

Sending Down Tail Block and Gantline.—Hold the weight of the gantline with one hand, or take a temporary turn with one part before letting go the block. The man on deck assisting should now form a large over-hand knot or any kind of hitch by using a bight of the gantline, in order to prevent the block hitting the deck when it runs down the gantline at speed. The man aloft then drops the bight to the deck.

Sennet.—Out of the many kinds of sennet it is possible to make, only two types are occasionally seen on board ship, the others are purely ornamental, and only used for fancy work.

Flat Sennet.—Three kinds of flat sennet are in general use, the only difference being the number of ends used. Almost any odd number of ends could be employed but three, five and seven would make sennet wide enough for all ordinary purposes. The method used is the same when plaiting with three ends: the outside end on the side containing most strands is brought to the centre, which means that each side alternately contributes a strand towards the centre. For instance, when working with five ends, three would lie on one side and two on the other. The outside one of the three would then be brought across to the smaller group of two, which would now of course contain three, and then the outside strand of this group would be taken across to the other side and so on.

Round Sennet.—Take six ends of equal length and whip them firmly at one end. Pick out three alternate ends and hold them up in a vertical position. This leaves an end hanging down between each of the vertical ends, and by continually interlocking these opposite numbers the round sennet is formed. Start with a vertical end and lead it over the right-hand end hanging down, then do the same with the two remaining vertical ends. Follow on in this manner until the required length is reached.

Coxcomb.—A simple and purely ornamental method of covering the eye of a rope. It is seldom used except by seamen when making fancy ditty-bag handles for their own use. Two ends may be worked but it will be found that three lies much neater. Always start at the middle of the eye and work down, or along each side. Material used is according to size of rope to be covered and is in the same proportion as serving a rope.

(a) *Two Ends.*—Take two lengths of small stuff and attach the middle part to the outside crown of the eye with a light temporary seizing. Keep the crown towards you, pull the hitch tight as you go, and work away from the person; when one side is finished, cut the temporary lashing and work along the other side.

Procedure.—Lay left-hand end across towards right side. Lead right-hand end across it to the left and form a half-hitch on the rope itself. Lead the hitched end back to the right again. Pass the left-hand end across it to the right and again form a half-hitch on the rope itself. Keep repeating this procedure until the desired length is obtained.

(b) *Three Ends.*—Start in the same manner as when two ends are used.

Procedure.—Pull right-hand end towards centre of rope. Lead centre end across it (towards right) and form a half-hitch on the rope itself. Pull left-hand end towards centre. Lead centre end across it (towards left) and again form a half-hitch on the rope itself. Repeat this procedure until one side of the eye is covered, cut temporary lashing and complete second side by working in opposite direction.

Setting up Gripe Lanyards.—"Ratline stuff" is generally used for setting up gripes, and the lanyard should be spliced to the link attached to the short length of chain, and not to the link attached to the deck.

Reeve towards the person. When sufficient turns have been taken, form a marline hitch with the hauling part and slip over the end of a handspike, but before taking a good heave apply a little grease to help the parts to render. When sufficiently tight the end should be passed through the middle of the lashing, wrapped tightly around all parts and finished off with a half-hitch, the end being stopped to the lashing itself with sail twine.

Setting up Stays.—Topmast backstays can usually be set up without much effort, but fore and main stays, on account of the greater angle generally require either steam power or a "handy billy", or both. Sometimes a single messenger is sufficient, if a suitable lead to a winch can be found, but this is seldom the case.

Slings, Making Up.—Timber stacked neatly for slinging is easier to strop and safest to hoist, but very often a load of timber is simply thrown in a heap, stropped, and when off the deck a variety of lengths will slide out of the sling, endangering men's lives who may be underneath. To prevent this, small pieces should be laid across the lengths of timber near the ends while it is being built up, in order that pressure will be communicated to the middle of the sling when the strop tightens.

A round turn should be taken around the whole load before reeving the end.

These remarks apply to all slings containing long poles, rods of metal and particularly to pipes.

Shortening Chain Slings.—Pull slack of chain through big ring to form a bight. Pass one hand through the bight, catch hold of standing part, let everything else drop and place standing part over the cargo hook ready for heaving.

Shortening Rope Slings (or Strops).—Make a simple overhand knot with two bights formed on the spare end of the strop itself. Hang both bights on the cargo hook. Fig. 52.

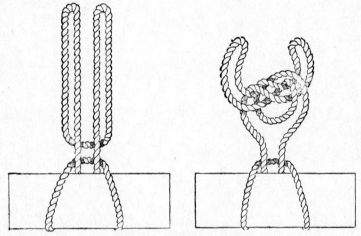

Fig. 52.—Shortening Strop.

Fitting a Tail to a Wire.—Sometimes the end of a wire attached to a towing "spring" is fitted with a long rope tail from 6 to 12 fathoms in length, and, in view of the fact that it may have to be renewed occasionally, the simplest way to fit it is to form a coxcomb by working the rope strands around the wire itself. Proceed as follows in the case of a 4-inch wire with a $3\frac{1}{4}$-inch rope tail.

Make a half hitch with the tail about 8 inches from the end of the wire and leave about 7 feet of end.

Pass a good, firm seizing on each side of the half-hitch, the first one about 3 inches from the end of the wire and the second one about 6 inches past the hitch.

Unlay 6 feet of end and begin coxcomb 12 inches from the hitch (6 inches from last seizing).

To form the coxcomb, arrange the ends so that they emerge from the underside of the rope and lead up towards the left.

Take the lowest (or the one farthest from the end), lead it over the other two and form a half-hitch on the wire by passing the end around in a clockwise direction.

The strand which was formerly the middle one will now be the lowest, and this is disposed of in the same manner by leading it over the other two and forming a half-hitch as before.

Keep repeating this procedure and, as the hitches accumulate along the wire, it will be found that the strands will lay up into the form of a rope with a little persuasion, but care must be taken to keep the hitches in a straight line along the wire or they will develop a spiral effect.

Make 10 hitches, then lay up the ends again for a distance of 12 inches, whip neatly and cut.

This end having been formed is secured to the wire with two seizings, one near the end of the rope and the other about 4 inches from the last hitch.

Another method preferred by some when fitting a tail to a wire is to proceed as before, starting off and finishing in the same way, but forming a different type of coxcomb, as under:—

With the strands lying in the same position as before,

Lay right-hand end towards centre of wire.

Lead centre end across it (towards right) and form a half-hitch on wire itself by passing the end around in an anti-clockwise direction.

Lead left-hand end towards centre.

Lay centre end across it (towards left) and again form a half-hitch on the wire itself by passing the end around in a clockwise direction.

Repeat the procedure for any required distance then finish off in the same manner as before.

Fitting Hose Couplings.—First of all it is necessary to "rig the job up" in a manner which will be most convenient for working. A vice is out of the question for holding the coupling firmly in position, seeing that it is of soft metal and easily damaged, so the best method is to lash a handspike in a horizontal position, at a suitable height, either to a rail or other convenient place, or it may be held in the jaws of a vice instead of lashing.

The coupling can now be slipped over the end of the handspike, then tapped lightly to jam it in position to keep it from turning, after which the hose is pulled into place over the coupling.

Before passing the seizing a strip of thin leather or a piece of canvas about $2\frac{1}{2}$ inches wide must be placed around the hose, in the exact place where the wire seizing will lie, in order to protect the surface of the hose itself from the wire, which will bite into it if put on firmly, as it should be.

It will be noticed that most new hoses when they come from the makers are seized with thin copper wire, but it is not likely that this will be obtainable on board the average vessel, so ordinary seizing wire will have to be used.

A single strand from a length of seizing wire would do very nicely, but this would mean spoiling about 25 feet of wire, so this method is seldom used. The usual practice is to take small gauge seizing wire, or, if only a heavy type is obtainable, halve it.

For an ordinary 5-inch hose about 8 feet of wire will be required.

Begin in the usual way by serving over the end of the wire which should be unusually long, say about 12 inches, and work towards the end of the hose.

Pass about five turns, then twist up the standing part with the shorter end (about six twists). Carry on serving over the short end again, passing another five turns, then finish off by twisting both parts together (about six twists). Cut and hammer the end down to prevent possible injury to the hands.

This method ensures that if any part of the seizing breaks, only one-half of it will come adrift.

Pulling Seizing Wire Tight.—To pull seizing wire tight is sometimes difficult. For one thing, it is small to hold, and to make matters worse, is often greasy. An ordinary hammer is the best remedy for this. Lay the handle (near the head) on the hauling part, make a round turn with the free end, and pull tight. In this way the free end is jammed against the head by the standing part, forming a sort of blackwall hitch, and a good pull can be taken on the wire without fear of it slipping—especially if the spare end is held in the hand.

Unlaying Seizing Wire.—It is often necessary to take one or more strands out of a long length of seizing wire. The best way to do this is to unlay about 600mm at one end, then hold a part loosely in each hand so that both parts are free to revolve.

Another man at the other end of the wire (which is stretched out to its full length) should then bend the end to make a small handle, and this he will turn with the right hand, while steadying it with the left. The "turns" communicated to the opposite end will quickly unlay the strands.

Strain on Hoses.—When taking fresh water from a shore hydrant with a hose rigged from ship to quay, the distance between may be considerable, thereby putting a good deal of strain on the hose. Use a heaving line to take the weight, passing it around the hose in long spirals.

Wracking.—This is a method of holding two ropes together when they move in opposite directions. A substantial piece of small stuff must be used if there is any great strain on the parts. First of all, pass a bight around both parts and reeve the ends through, so as to pull both parts together, then pass the wracking around both parts and between them with each turn, in figure-of-eight fashion. Sometimes the two ends are separated and passed around in the same manner, but each in opposite directions. Whichever method is adopted the effect is the same, and as a wracking is usually just a temporary measure, the ends are finished off with two half-hitches. The number of turns depends upon the amount of strain or pull on the parts wracked off.

Wracking Off a Fall.—Sometimes it is not possible to use a stopper on a fall, in which case a wracking seizing is applied to any two "parts" which are moving in *opposite* directions.

CHAPTER IV.

DECK STORES AND GEAR

Accommodation Ladder. Blocks. Shackles. Tackles. Purchases. Thimbles. Hooks. Cargo Gear. Hatches. Derricks.

ACCOMMODATION LADDER.

Bridle.—An iron crosspiece on the end of the accommodation ladder tackle, to each end of which is attached two chain legs which are shackled to the ladder itself. The principal function of a bridle, apart from holding the weight of the ladder, is to keep the legs the same distance apart as the width of the ladder, so as to allow a clear passage in between them.

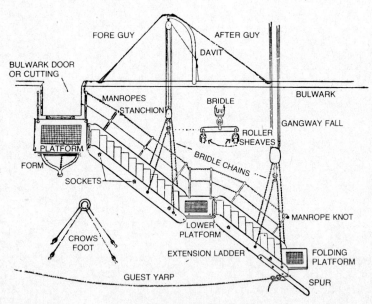

FIG. 1.—Accommodation Ladder.

Platform.—A hinged platform which, when lowered level, is supported underneath by a fork slotted to the ship's side. The accommodation ladder is hinged to the platform.

Lower Platform.—On the larger vessels it is necessary to fit an extension-piece to the accommodation ladder, in order that it will reach to the water's edge, and when this is done another platform is fitted to the bottom part of the top ladder, and the extension is connected to this lower platform. Fig. 1.

Folding Platform.—This is the folding step or platform near the water's edge, and from it passengers enter or leave boats lying alongside.

Stanchions.—Steel posts fitting into lugs on each side of the gangway or accommodation ladder; through holes in the top ends of the stanchions manropes are rove to act as guardrails.

Manropes.—Ropes rove through stanchions on the accommodation ladder. They are usually covered with canvas and are always finished off with a manrope knot at one end.

Crow's Feet.—A substantial steel ring with four chain legs shackled to the four corners of the lower platform, for lowering it down to, or heaving it up from, the sockets on the ship's side in which it rests.

Screens.—Canvas screens lashed to the handrails and steps, on both the outboard and inboard sides of the accommodation ladder.

Spur (or Horn).—To prevent small craft getting jammed underneath an accommodation ladder, it is usual to bolt a long piece of wood to the outboard side, in such a way that it extends well down into the water.

BLOCKS.

Construction of Blocks.—Oak and elm have such good weather resisting qualities, and stand such a lot of wear and tear, that most blocks are made from these materials. Blocks are measured by the length of the shell, and sheaves by their diameter. The various parts are as follows:—

Shell.—The outside case.

Fig. 2.—Sheaves.

Sheave.—The grooved wheel over which the rope travels. They are made of either phosphor bronze, steel or lignum vitae, according to the type of block to which it is fitted. Lignum vitae is an exceedingly hard wood with self-lubricating qualities, and it makes the best sheave of any. Sheaves are measured by the diameter.

Roller Sheave.—An ordinary sheave with metal rollers fitted to the bush which work round the sheave pin. Ball bearings are often used for the same purpose.

Bush.—The metal centre of a sheave through which the pin passes. In large gin blocks an oil reservoir is fitted round the bush.

Sheave Pin.—A steel axle pin passing through the shell and sheave, on which the sheave revolves.

Score.—The groove on the outside shell of a wooden block to take a wire strop. The score is deepest at the bottom or tail of the block to take the splice.

Crown.—The top of the block.

Tail.—Bottom of the block.

Swallow.—An opening between the cheeks of the shell through which the rope passes.

Cheeks.—The sides of the shell on each side of the swallow.

Clump Blocks.—These are cut from the solid piece of wood and are very rounded in shape. They are seldom seen nowadays.

Made or Built Blocks.—As the name implies, these are built from separate pieces cut from the plank. The cheeks are riveted together with a piece of wood between them at each end to form the swallow, and into this the sheave is fitted.

Steel Bound Blocks.—A built block which, in place of a wire strop, is fitted with a heavy steel strop or band. This runs round the outside of the shell to give it added strength. The sheave pin goes through both strop and shell. A hook or eye is fitted to the steel strop at the crown of the block, and sometimes a small one at the tail also, so that the standing part of a tackle can be shackled to it.

Internal Steel Bound Block.—A popular type of built block with a removable fork-shaped strop inserted down through each side of the shell. One of the forks is usually longer than the other, so that it may extend through the bottom of the block and form a lug, to which is shackled the standing part of the fall. Holes in each fork coincide with others in the shell to take the sheave pin.

Non-Toppling Block.—A special type of block used mostly for lifeboat falls. The tail is the heaviest part, and this permits a fall with no load on it to be "rounded up" from the water's edge without toppling over or becoming jammed. Figs. 5, 6.

Snatch Blocks.—Small strong steel blocks with a hinged side. This permits a fall to be put over the sheave through an opening in the side, without reeving the end through. Used as lead blocks they are very useful when moving cargo on deck or when lashing cargo. Larger sizes will take mooring ropes.

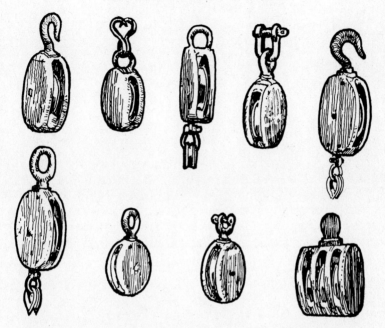

FIG. 3.—Several Types of Wood Blocks with Different Fittings.

Lead Blocks.—Any type of block may be utilised as a lead block, but a proper one is always fitted when gear is rigged permanently. For instance, when the hauling part of a topping lift fall has to be led to a winch, the fall will be found rove through a lead block shackled to the deck at a convenient angle.

Also, when a heavy lift derrick is working, the guy falls have usually to be led through one or more lead blocks to a distant winch.

FIG. 4.—Types of Steel Bound Blocks.

They are made of steel, and have either a swivel eye or hook fitted to the crown, and as the sheaves are wider than cargo blocks of the same size, a 12-inch lead block could, for instance, be used along with a 14-inch cargo block and take the same sized wire. The complete range of sizes in lead blocks is: 6, 8, 10, 12 and 14-inch.

FIG. 5.—Ordinary Type Welin   FIG. 6.—Welin D.C. Type
Non-toppling Lower Block.   Non-toppling Lower Block.

FIG. 7.—Types of Metal Blocks.

Tail Block.—It is used principally aloft when a gantline has to be rove off for any purpose. A small wooden single sheave block with a long tail of rope (about one fathom), will be found best. Fig. 9.

Funnel Blocks.—A number of small steel blocks may be found shackled around the top rim of the funnel. Gantlines are rove through them for either funnel boards or bo'sun's chairs when painting or repair work is in progress.

Hook Blocks.—Wooden blocks which, instead of a shackle or strop, are fitted with a big iron hook so that they may be hung on the rim of a funnel. They are meant to take the place of metal funnel blocks as these are not always safe after a few years aloft without proper attention. One big advantage is that they may be hung in any required position. Fig. 10.

Butterfly Block.—A small snatch block cut out from a solid piece of wood, with a short tail attached. They are seldom seen on modern ships, but were used at one time for hauling in the deep sea lead.

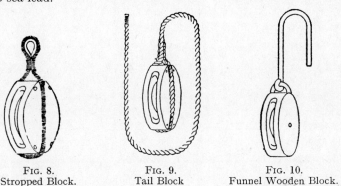

| FIG. 8. | FIG. 9. | FIG. 10. |
| Stropped Block. | Tail Block | Funnel Wooden Block. |

Cargo Gins.—Two cargo gins are fitted to each derrick; one at the head, which is variously known as a head block, gin block, or cargo block, and the other at the bottom known as a heel block. They are made of steel and come in for much wear and tear, requiring to be frequently oiled and overhauled. They have self-lubricating sheaves and a swivel eye. Fig. 11.

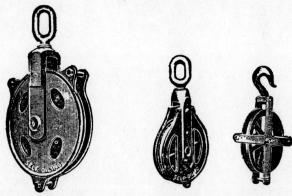

FIG. 11.—Cargo Gin.   FIG. 12.—Coaling Gins.

The diameter of the sheave is stamped on the shell of each block, and it is known by that size, the full range of which is 6, 8 10, 12, 14, 16 and 18-inch.

Heel Blocks.—The only difference between a heel block and a cargo gin is the method by which they are fitted into place. Instead of an eye, they have a small swivel attachment which keeps them at right angles to the "table". Also, the sheaves are wider than the same sized cargo gin, but are stamped with the diameter of the sheave on the shell, and it is known by that size. The range of sizes is 150mm to 360mm.

Coaling Gins.—These are made with a steel frame, with self-lubricating sheave and swivel hook or eye, and are used chiefly for working coal. Like cargo gins they are known by the diameter of the sheave, the full range of sizes is from 150mm to 400mm. Fig. 12.

Overhauling Blocks.—All blocks must be overhauled periodically, the time limit depending on the type of block and the amount of work it performs.

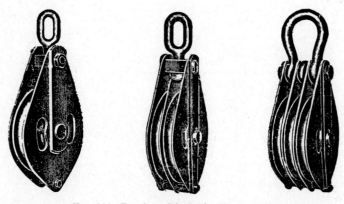

FIG. 13.—Purchase Blocks for Heavy Lifts.

Wooden Blocks.—The sheave pin must be punched out with a spike or other sharp instrument, and all working parts greased. Solidified vaseline is often used for this purpose in place of blacklead and tallow.

Steel Gin Blocks.—Cargo gin blocks on the derricks come in for the most attention, and are frequently overhauled.

When the sheave pin is removed and the sheaves taken out, they are first scraped and then coated with a mixture of blacklead and tallow. The inside cheeks of the gin should also be coated with the mixture. On the sheave itself will be found a small screw acting as a plug to the oil reservoir. This feeds the sheave pin through the bush. Remove the pin and fill the reservoir.

While on the job it is as well to remove and oil the distance piece, if this has not already been done, for the purpose of taking out the sheave.

The swivel at the head of the block will probably be quite stiff to move, or even "frozen up" altogether, and it might require liberal doses of paraffin to free it, before applying lubricating oil.

SHACKLES.

Shackles.—Sizes and types cover a wide variety, ranging from about 1 inch in length to a size capable of lifting many tons. They are made of wrought-iron or cast steel and are used for all joining purposes.

For some classes of work, such as standing and running rigging and the handling of cargo, only tested shackles are used. On these the "safe working load" in tons will be found stamped into the iron, thus: S.W.L. (5) T.

Shackles smaller than $\frac{3}{8}$ inch are never tested and those in use for cargo work range from about 5 to 10 tons. Anything heavier than that would be for some special purpose, such as the heavy lifting derrick, mooring wires or oil rig anchors.

Harp or Bow Shackles.—Along with D shackles these are the two principal types to be seen on all vessels.

    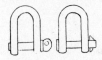

FIG. 14. Harp Shackles   FIG. 15. "D" Shackles   FIG. 16. Awning Shackles

D Shackles.—One of the two principal types in general use. Fig. 15.

Mooring Shackles.—A large, heavy type with a bow handle to the pin will be found on most ships. They are used almost exclusively for attaching mooring wires to the ring on a buoy or wharf, and also for heavy lift work, when the sling or strops around the "lift" (or weight) have to be shackled to the purchase block of the heavy lifting derrick. Fig. 17.

Pins.—Instead of the usual screw type, some shackles are fitted with an easy-fitting pin which is kept in place with a forelock.

Others again have either a square socket, or a slot on the pin similar to that of a screw head, in which case a special shackle key is required to screw the pin out. This type is often used in connecting up cargo runners in order that the projecting end of a shackle pin, which is inclined to catch on other objects, is dispensed with. 35-50 and 75 ton shackles often have a nut and bolt as a pin. A splitpin stops the nut working back.

Fluted Shackles.—On cable-laying vessels about 15 types of shackle are in constant use, and although of different sizes and types, they all have one thing in common as regards the pins. These are slotted for their full length and coincide with a slot on the lug of the shackle, which enables a wire to be passed through to lock the pin.

Anchor Shackles.—Large heavy shackles which join the anchor to the cable. They are heavier than the shackles which join the different lengths of cable together, but the pins (which do not project beyond the greatest width of the shackle), are held in place with a wooden or metal plug which is hammered through the lugs and pin also. Fig. 18.

Cable Shackles.—These are frequently of the patent lugless type.

Awning Shackles.—Ordinary D type with a thimble set at right angles to the crown of the shackle. These are used for shackling an awning to a ridge wire, and the thimble permits another awning to be made fast to the same shackles with rope "stops". This type is only employed when ridge wires are fitted in place of wooden spars. Fig. 16.

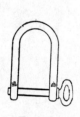
Fig. 17.
Mooring Shackles.

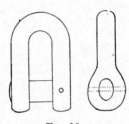
Fig. 18.
Anchor Shackles.

Fig. 19.
Lugless Shackle.

TACKLES AND PURCHASES.

Tackle.—A simple device which gives increased power by a combination of blocks and ropes. The number of sheaves in the block, the manner in which the rope or fall is rove through them, and whether or not the standing part is made fast to the top or bottom block are all distinguishing features of the various types of tackle. The theoretical power gained is proportionate with the number of sheaves in the tackle and varies from two to nine times according to the type of tackle used.

Top Block.—The one through which the hauling part runs is invariably the top block.

Bottom Block.—The opposite block to that having the hauling part, and the one which is usually shackled to the deck.

Standing Part.—The part made fast to the block.

Hauling Part.—Another name for the fall, or the part which is manned when heaving.

Two Blocks.—When blocks are hove close together they are said to be "two blocks".

Overhauling.—The act of stretching blocks farther apart by pulling in the opposite direction to that taken when heaving.

Choked Luff.—A term which indicates that one part of a fall has jammed underneath its neighbour, therefore preventing further movement.

Thoroughfoot.—Sometimes the opposite block to that through which the hauling part is rove becomes accidentally dipped through the fall. This causes the parts to cross and is known as a thoroughfoot. To clear it the block must be dipped back again in the opposite direction to which the turns take.

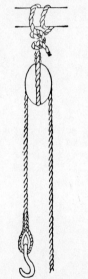

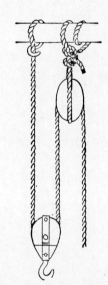

FIG. 20.—Single Whip.   FIG. 21.—Double Whip.

Rounding Up.—Small tackles are stowed away "two blocks" with the fall coiled neatly on top. Pulling the slack through until the blocks come together is known as "rounding up the fall".

Lifting Power of Tackles.—Take the weight that a single part of rope is capable of suspending and multiply it by the number of parts at the movable block. This will give the theoretical weight the tackle could lift but roughly one-tenth the weight must be deducted for friction at each sheave so that in the case of a luff tackle roughly one-fourth must be subtracted to give the actual weight the tackle could safely lift.

Single Whip.—This is simply a rope rove through a tail-block with a single sheave. No power is gained, it merely acts as a lead block. Fig. 20.

Double Whip.—To "double up" a single whip, reeve the standing part through a hook block and make the end fast alongside the tail block. The power gained is two times. Fig. 21.

Gun Tackle.—Two single hook blocks. Power gained is twice or three times. Fig. 22.

Handy Billy or Jigger.—A double block with a tail, and a single hook block to which the standing part is made fast. This gives a power gain of three or four times. Fig. 23.

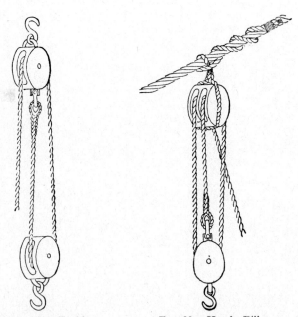

Fig. 22.—Gun Tackle.   Fig. 23.—Handy Billy.

Watch or Luff Tackle.—The same as a handy billy, but having two hook blocks instead of one tail block and a hook block. Power gained is three or four times. It is used for all purposes where an extra pull in excess of the man-power available is required. Fig. 24.

Double Luff Tackle.—Two double hook blocks with a power gain of four or five times. Fig. 25.

Three-fold Purchase.—This gives a power gain of six or seven times, and is the type of purchase used for the lowering and hoisting of lifeboats. When used for this purpose on a davit, it will be found that the fall is invariably led through the centre sheave.

Reeving Three-fold Purchase.—Lay the top block A with one cheek to the deck, and make sure it is the side with the becket. Next, lay the bottom block B also on the deck, with the swallows up and down, and with the bottom of both blocks facing one another.

First, reeve through the middle sheave of block A from right to left, down through the left-hand sheave of block B, through the bottom sheave of A from left to right, up through the right-hand sheave of B, through the top (or uppermost) sheave of A from right to left, down through the middle sheave of B and thence up to the becket where it will be shackled on after an eye has been spliced in the end with a thimble as indicated by the numbered arrows on the parts of the fall below.

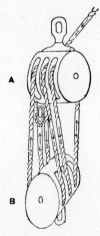

Three-fold Purchase (Rove through Centre).

Four-fold Purchase.—Four-fold blocks (and over) are only used on "Jumbo" derricks for very heavy lifts, the power gained being eight or nine times.

Gyn Tackle (or three and two tackle).—The latter name is perhaps more generally used than the correct one, probably because it is easier to remember the type of tackle required if the number of sheaves at each end are mentioned. In fact, it would be an advantage if all tackles were described in a like manner. Fig. 25

This type is used extensively as a Topping Lift Tackle.

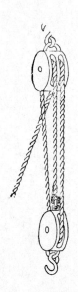

FIG. 24.
Watch or Luff Tackle.

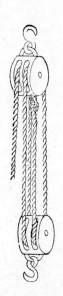

FIG. 25.
Double Luff Tackle.

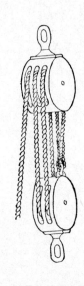

FIG. 26.
Three and two Tackle.

THIMBLES.

Thimbles.—There is no standard method of distinguishing or even naming the various sizes and shapes of thimbles; consequently, some people measure by length, and others by width. Davey & Co. Ltd., the London manufacturers, state that they consider the best method is to measure by width, for the reason that the width of a thimble is the same as the circumference of the rope which it will take. For instance, a thimble 1½-inch wide has a score which takes a rope of 1½-inch circumference. In metric terms the internal width of a thimble is approximately three times the nominal diameter of the rope it will take in its score, a 70mm thimble will take a 26mm rope.

Thimbles in general use are made from galvanised steel, but they will also be found in brass and gunmetal.

Sizes range from less than ½ inch to 12 inches, which is the largest type likely to be found afloat. Fig. 27.

Heart Thimbles.—This is the type which is used for practically all purposes on board, and other varieties are mostly an adaptation of this to suit special cases. For instance, some will be found to be extra long, while others may have extra width.

Lanyard Thimbles.—These are really heart thimbles broadened out at one end until they are triangular in shape. This allows many more turns of a lanyard to be passed through the thimble than usual.

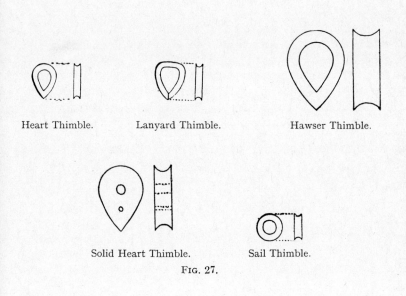

Heart Thimble.   Lanyard Thimble.   Hawser Thimble.

Solid Heart Thimble.   Sail Thimble.

Fig. 27.

Hawser Thimbles.—Extra wide and strong steel heart thimbles designed for heavy hawsers.

Solid Heart Thimbles.—These are sometimes called "deadeyes", probably because they take the place of the old-fashioned wooden "deadeye" on the shrouds.

They are usually of galvanised steel, with a large hole near the centre of the wide end. To this a bottle screw is attached when setting up the shrouds. At the narrow end another hole takes the bolt from the sheerpole.

Sail Thimbles.—In order that they will fit more snugly to canvas, all sail thimbles are made round, and may be of galvanised material, gunmetal, or brass. The former will always be found in awnings, while the two latter are only found on yachts.

HOOKS.

Beam Hooks.—These are used for lifting out the heavy steel beams fitted inside hatch coamings. They usually take the form of a substantial iron ring (Ring Clew) with two wire legs spliced into it, and these are attached to the beams by means of a hook on each leg.

Note.—Shackles may also be used in place of hooks.

Can Hooks.—*See* Cargo Gear.

 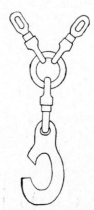

FIG. 28.—Cargo Hook.   FIG. 29.—Union Hook.

Cargo Hooks.—A very heavy hook with a swivel ring, which is shackled on the end of a cargo runner when heaving cargo out of holds. When two runners are used (Union Purchase Method), they are each shackled separately to a short length of chain with a swivel on what is known as a Union Hook. The swivels prevent turns or kinks forming in the runners. Fig. 28 and 29.

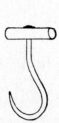 

FIG. 30.—Stevedore's Hand Hook   FIG. 31.—Chain Hook.

Chain Hooks.—Heavy chain such as anchor cable is very awkward to handle so long hooks about 1 metre in length are always provided for the purpose. They will always be found near the chain locker. Fig. 31.

Sister Hooks.—If, through shaking or flapping about (particularly with sails), an ordinary hook is liable to become disengaged, sister hooks may be used with safety. They are made from wrought iron, with short, wide heart thimbles to take hemp rope. A mousing gives added security.

FIG. 32.—Sister Hooks.

Hook Block.—*See* Blocks.

Hook Chair.—A Bo'sun's chair with a substantial hook seized to the strop, and used principally when painting samson-posts and fish-plates. When hooked in a convenient position, it takes the weight off the legs and arms, and allows free play with the hands.

Hook Rope.—A length of line with any kind of hook attached may be described as a hook rope. As the name implies, it is used for hooking anything up from overside.

Plug Hatch Hooks.—These are only found on vessels having refrigerated cargo space, and are used for lifting plug or insulated hatches. Although very much smaller, they are similar in design to beam hooks. Fig. 33.

FIG. 33.—Plug Hatch Hooks.

FIG. 34.—Tackle Hook.

Pot Hooks.—An ordinary hook shaped like the letter S, and used for hanging pots of paint in a convenient position.

Sail Hooks.—Small, sharp-pointed hook with a swivel eye attached to a lanyard, and made fast to a sailmaker's stool. It holds the canvas in place while the sailmaker is sewing.

Tackle Hooks (with Thimble).—Sizes are governed by the length of the hook, which is made from forged iron. The thimble is usually made short and wide to take hemp rope. Fig. 34.

Union Hooks.—Heavy cargo hook with swivel ring to which two short legs of chain are fitted.

A cargo runner is shackled to each leg when handling cargo by the "Union Purchase" method. Fig. 29.

CARGO GEAR.

Coaling Baskets.—Rattan cane from the Dutch East Indies is generally employed for the weaving of cargo baskets found on merchant vessels. At one time these baskets were used for handling all classes of cargo, but modern appliances have supplanted them to such an extent that they are now used almost exclusively for heaving rubbish out of holds, or working coal.

The sizes in general use are:—1, 2, 3 and 5 cwts., but 10 and 20 cwt. sizes can also be obtained if required.

Trays.—Some types of cargo are discharged with heavy wooden trays, which are lifted with a four-legged bridle, hooking on to a lug in each corner.

Dunnage Mats.—Straw mats used for the protection of cargo.

Dunnage Wood.—A cheap line of timber used for many purposes in the stowage of cargo, such as the protection of cases, levelling off uneven tiers, and ventilation. The different sizes are known simply by their thickness. For instance, 50mm × 50mm, 75mm × 75mm, and 150mm × 25mm are among the sizes generally met with.

Butter and Egg Battens.—Small size pieces of dunnage wood. Both types are made specially for use when stowing these commodities.

Quoins.—Triangular shaped blocks of wood. They are placed under the bilges or ends of casks to prevent them rolling about, and keep the bilge "free".

Cargo Runners.—Flexible wire rove through the head and heel blocks of a derrick, and attached to a winch for heaving cargo out of holds. The length varies according to requirements.

Whip.—Rope fall rove through a gin and used for "whipping" out cargo.

Scotsman.—A metal tube which fits over the fore or main stay to protect it from the chafe of the cargo runner.

Cargo Net.—Cargo which is too small to "sling" in a strop is thrown into a cargo net. This is a large rope net fitted with strops for attaching it to the cargo runner.

Save All.—Designed to answer the same purpose as a cargo net. As it is constructed with canvas, instead of netting, it will not allow anything to fall through, therefore it "saves all".

Snorters (or Snotters).—A single-piece sling, up to a couple of fathoms in length, with an eye at each end. They are used for "slinging" cargo.

Wool Snorters.—A bunch of four snorters knotted together at one end, so as to leave four legs for lifting four bales of wool at the same time.

Strops.—The two ends of any single piece of rope spliced together will form a strop. Cargo is "slung" with these strops, and they are used for many other purposes, especially when a tackle or hook has to be attached to any object.

Wire Snorters (or Slings).—These are made in the same way as a rope snorter and perform the same function, particularly where machinery is concerned.

Wire Strops.—Are used mostly for slinging machinery, or anything else liable to cut through a rope strop.

Chain Slings.—Many lengths and sizes of chain slings will be found. They come in useful for slinging machinery or as lashings.

There is a large egg link at each end—one larger than the other—so that one end may be rove through the other. Sometimes a hook is fitted to one end instead of a link.

Collar Slings.—Length of chain with an egg link at each end.

Single Slings.—Length of chain with a ring at one end and a hook at the other.

Double Slings.—Large ring with two lengths of chain having a hook on each.

Can Hooks.—A simple and effective means of lifting heavy casks. Two broad hooks, running free on the bight of a rope or chain sling, are hooked under the chime of the cask at each end. They are quickly adjusted to suit any size of cask, the weight of which holds the hooks in position. Fig. 35.

Casks containing liquid are usually slung with a rope strop.

Pallet.—Many cargoes are now strapped to a pallet for ease of handling by fork-lift trucks. The double wooden base means that no dunnage is required.

Stevedores' Hand Hook.—Small hand hook used by stevedores when handling bales and boxes. Fig. 30.

Slinging Barrel on End.—Place the single bight of a strop across and under the bottom of a barrel standing on end. Throw a half-hitch over the top half with each part, so that the two standing parts will be directly opposite one another. This will leave a strop above the barrel which can be hooked to the cargo runner. Fig. 36.

FIG. 35.—Can Hooks.

Another method is to take an ordinary end of rope, place a bight under the lower end and make an overhand knot across the top. Open out the overhand knot or crossed parts until big enough to slip over the top of the barrel. Adjust and set tight, and make the short end fast to the hauling part, at a suitable height above the barrel, with a bowline.

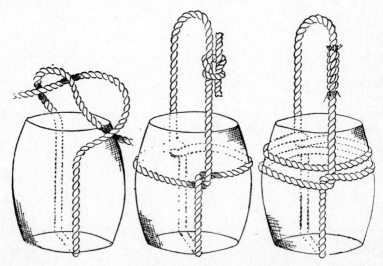

FIG. 36.—Slinging Barrel on End.

HATCHES.

Battening Down Hatches.—To ship the hatch beams, putting on the wooden hatch covers, spreading the tarpaulins and wedging them to the coamings, is known as "covering up" or "battening down".

Hatch Beams (British Standard 4263).—Beams must be shipped in their proper places or the wooden hatch covers will not fit properly. Each one has number stamped on the steel flange at one end, and also letters indicating whether it is the end which goes to port or starboard.

As a rule these particulars are painted on the beam to make them more prominent, and in addition, a large green and red patch indicates the respective ends.

No. 1 beam is usually the forward one in each hatch.

Wooden Hatch Covers (British Standard 4268).—For identification each hatch board is numbered with a figure cut into the wood but in the old days they were seldom put on in their proper order mostly because the ends became so badly worn that they were put on where they would best fit. However, since 1946 it has been compulsory under Board of Trade Rules to fit galvanised protections to the ends.

The hatch boards are usually about 9 feet in length, $2\frac{1}{2}$ inches in thickness and 18 inches in width, being made up of two 9-inch boards laid side by side and joined near each end by iron rods passing through holes drilled through the two boards from side to side. The ends are strengthened by 3-inch galvanised iron bands and handles are formed by gouging out hemispherical holes, an iron bar being screwed across the recess.

Distinguishing Colours for Hatches.—Insulated hatch beams and plug hatches fitted to a number of decks will often become hopelessly mixed after a long stay in port.

To prevent this, and to facilitate "shipping them", it is a good plan to use a different colour for the plugs, beams and coamings of each deck, as under:—

Upper 'tween deck	Red.
Middle 'tween deck	White.
Lower 'tween deck	Black.
Lower hatch deck	Other.

Spreading Tarpaulins.—Tarpaulins are made from a special tarpaulin canvas, and are sometimes given a coat of Stockholm tar to make them more weatherproof—especially on the Western Ocean route. The particular cover for each hatch is easily determined by examining the four corners.

One or more of them will have a number of small eyelets stamped in to coincide with the number of the hatch. Another guide to the spreading of covers is the seams. Sometimes they run fore-and-aft, and sometimes athwartship. Once you have acquired this information it will always be a guide on that particular ship. When the seams run athwartships it should be remembered that, like an awning the loose end of a seam is always the after part, so that it will not catch any water sweeping over it from forward.

The ship's carpenter usually supervises the spreading, tucking in and wedging of the tarpaulins, and the nature of the voyage, as well as probable weather conditions, will determine the number and order of the tarpaulins to be used.

It would be safe to say that a deep loaded vessel invariably has three tarpaulins to each hatch, while a vessel which is "light", only has one, or at most, two.

When three are used, orders may be given that the newest one will go on top, with the old ones below.

Tucking In.—The trouble when three tarpaulins are used is that they will not all fit snugly in the cleats; therefore, it is a common practice to fold either the first or second one (usually the oldest) in at the edge of the coamings.

Two tarpaulins may be tucked in together, but it is better to tuck each one separately, so that, at the corners, the canvas will be more evenly distributed.

At the corners, the tarpaulin must be folded away from the direction in which heavy seas are likely to come across the deck.

Shipping the Battens.—The "turn up" of a cover should just touch the bottom of a cleat. If the batten is allowed to rest on a bight of canvas, damage to the canvas will be the result.

Wedges.—These are made from tough woods such as elm which weathers well and tends to swell and grip tighter when they become wet in bad weather. They should be placed in the cleats with the longest side next to the coamings, otherwise the sharp edges of the cleats would cut into the grain of the wood when hammered tight.

Locking Bars.—When battened down, steel bars or equivalent must be provided to secure each section of the hatch covers. They lock in place and also prevent unauthorised entry.

Stripping Hatches.—Before cargo can be worked through a hatchway, the tarpaulins, hatch covers and hatch beams must be taken off. This is known as "stripping hatches".

When taking off hatch covers, begin in the centre line of the middle section and work out to the sides of the hatch. The covers should be stacked neatly near the coamings, and should not be thrown about too much or damage will result.

Hatch beams landed on the deck invariably require dunnage wood underneath them, either to protect the deck if a wooden one, or to facilitate casting the beam slings adrift.

Tarpaulins may be made up in any manner which suits a particular hatch, and should be thoroughly dry if they are to be stowed away for a lengthy period. A rope yarn should be attached with a number of knots corresponding to the number of the hatch. This will make their identity much easier.

Watertight Steel Hatches.—As a result of the International Conference on Load Lines, 1966, vessels fitted with wooden hatch covers are by the New Load Line Regulations now in force, not allowed to load as deeply as vessels fitted with steel hatch covers. Though not obsolete the wooden hatch covers are rapidly being superseded by steel sectional slab hatches such as have been perfected by MacGregor & Co., Ltd.

By combining the beams and covers, great strength has been achieved and power-operated mechanism which may be hydraulic or by wire and winch, gives an easy and quick means of sealing the hold with a safe watertight cover whether of the single pull or folding type. All the steel cover sections roll on wheels in a channel on the coaming and pivot at the end of the coaming into a vertical stowage position against the winch island or masthouse bulkhead.

In the closed position the steel covers are secured by screw cleats or by clam locking cleats fitted at intervals round the coaming and when tightened these compress the rubber jointing on which the cover rests to form a watertight seal. The hatch can just as quickly be opened up either fully or in part and as the covers are self-stowing there is a considerable saving of time and labour whilst the decks are left comparatively clear for working.

It is essential to see that the cover wheels are kept greased and free from grit and that the coaming and drainage channels are kept clean. Securing cleats must be kept greased and should be checked for tightness when at sea. (*See* Figs. 37, 37a.)

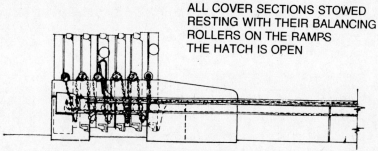

ALL COVER SECTIONS STOWED
RESTING WITH THEIR BALANCING
ROLLERS ON THE RAMPS
THE HATCH IS OPEN

Fig. 37.—Operation.

Dangerous Cargo.—Acids, corrosive fluids, or anything of an inflammable nature is not allowed to be stowed down a hold. It is kept on deck, so that in the event of leakage it is easily disposed of.

DERRICKS AND GEAR.

Derricks.—Derricks are also known as cargo booms and are provided solely for the purpose of loading and discharging cargo or stores so that the number to each hatch will depend upon the class of ship or the trade in which she is engaged. The original type

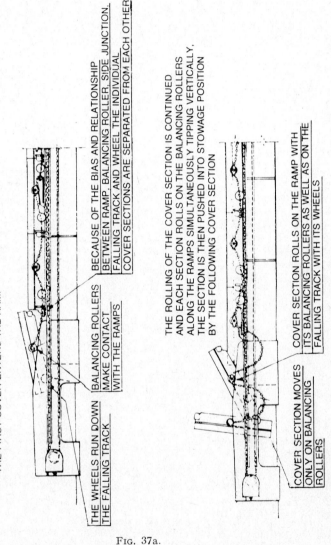

MACGREGOR STEEL HATCH COVER

THE LAST COVER SECTION IS ENGAGED TO THE MANOEUVRING CHAIN
THE WINCH IS STARTED AND RUNNING IN THE OPENING DIRECTION
THE FIRST COVER ENTERS THE RAMP

BECAUSE OF THE BIAS AND RELATIONSHIP
BETWEEN RAMP, BALANCING ROLLER, SIDE JUNCTION,
FALLING TRACK AND WHEEL THE INDIVIDUAL
COVER SECTIONS ARE SEPARATED FROM EACH OTHER

BALANCING ROLLERS
MAKE CONTACT
WITH THE RAMPS

THE WHEELS RUN DOWN
THE FALLING TRACK

THE ROLLING OF THE COVER SECTION IS CONTINUED
AND EACH SECTION ROLLS ON THE BALANCING ROLLERS
ALONG THE RAMPS SIMULTANEOUSLY TIPPING VERTICALLY,
THE SECTION IS THEN PUSHED INTO STOWAGE POSITION
BY THE FOLLOWING COVER SECTION

COVER SECTION ROLLS ON THE RAMP WITH
ITS BALANCING ROLLERS AS WELL AS ON THE
FALLING TRACK WITH ITS WHEELS

COVER SECTION MOVES
ONLY ON BALANCING
ROLLERS

Fig. 37a.

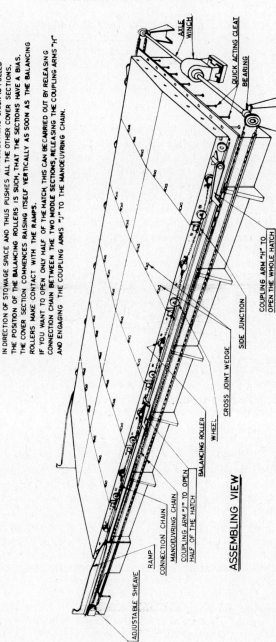

CROSS JOINT WEDGES AND QUICK ACTING CLEAT'S ARE UNFASTENED. COVER
SECTIONS ARE LIFTED AND REST ON THE WHEELS
HATCH IS NOW RAIN TIGHT AND COVERS READY TO BE ROLLED.
THE LAST COVER SECTION WHICH IS COUPLED TO THE MANOEUVRING CHAIN IS PULLED
IN DIRECTION OF STOWAGE SPACE AND THUS PUSHES ALL THE OTHER COVER SECTIONS.
THE POSITION OF THE BALANCING ROLLERS IS SUCH, THAT THE SECTIONS HAVE A BIAS.
THE COVER SECTION COMMENCES RAISING ITSELF VERTICALLY AS SOON AS THE BALANCING
ROLLERS MAKE CONTACT WITH THE RAMPS.
IF YOU WANT TO OPEN ONLY HALF OF THE HATCH, THIS CAN BE CARRIED OUT BY RELEASING
CONNECTION CHAIN BETWEEN THE TWO MIDDLE SECTIONS, RELEASING THE COUPLING ARMS "H"
AND ENGAGING THE COUPLING ARMS "J" TO THE MANOEUVRING CHAIN.

AXLE
WINCH
QUICK ACTING CLEAT
BEARING

COUPLING ARM "H" TO
OPEN THE WHOLE HATCH

SIDE JUNCTION

CROSS JOINT WEDGE

WHEEL
BALANCING ROLLER
COUPLING ARM "J" TO OPEN
HALF OF THE HATCH
MANOEUVRING CHAIN
CONNECTION CHAIN
RAMP

ADJUSTABLE SHEAVE

ASSEMBLING VIEW

Fig. 38.—MacGregor (Single Pull) Steel Hatch Covers.

E

known as the Mannesman Tubular Derrick is made up of seamless steel tubes or pipes, stepped one within the other so that the end diameter is two-thirds the diameter at mid-length.

The top of the derrick is known as the head and about 12 inches from below the top a metal collar or spider band is fitted carrying equispaced lugs. The two side ones are for the guy pennants, whilst the upper one takes the span or the topping-lift block and the underneath one takes the cargo gin. The lower end of the derrick is called the heel and is held in position by a gooseneck fitting which fits into a swivelling collar attached to the masthouse or samsonpost Immediately below the collar is a swivelling lug which takes the heel block, through which the cargo runner passes.

The S.W.L. (Safe Working Load) of a derrick must be painted or stencilled on it and all the gear used such as the cargo hook, shackles and blocks must each have its S.W.L. stamped on it. In some cases the derrick may have two sets of figures in which case the second one, often indicated by S.W.L.(U), refers to the Union Purchase Method of working derricks and meets the requirements of the Australian Safety Regulations.

In most dry-cargo vessels there are a minimum of two derricks per hatch with at least one heavy derrick (or "Jumbo" derrick) at the main hatch for heavy lifts, whilst in tankers with only liquid cargoes to cope with they are limited to two amidships, one each side to handle the pipelines from the shore and another aft for taking in stores.

Union Purchase.—This is the most popular method of handling cargo with two derricks in which one derrick plumbs the square of the hatch and the other plumbs the wharf. The cargo runners are joined together by means of a union hook. This allows a sling of cargo to be hoisted out of the hold by means of the first derrick and hove over the ship's side between the two ready to be finally lowered from the second derrick. (See Fig. 29, p. 110.) This method has already been referred to in the previous paragraph.

Single Derrick.—To handle cargo with a single derrick is a laborious system of continually heaving the derrick inboard and outboard, unless there happens to be sufficient winches to rig two steam guys.

Swinging Derricks.—With two derricks available, another fairly fast method of working cargo is to rig a swinging derrick on the discharging side, and leading an inboard guy to the head of the other derrick which is kept stationary on the opposite side of the hatchway.

Instead of an inboard guy on the swinging derrick, a wire runner is shackled in its place and rove through the cargo block on the stationary one. To this, at a suitable height, weights heavy enough to pull the derrick inboard are attached. (The weights are known as the "Dead Man".) In place of the usual outboard guy, a single part wire steam guy is fitted, and this will heave the derrick outboard, from which position it will be pulled back again by the weights on the stationary derrick, if both have been adjusted to suitable positions.

Anything, such as coils of wire and cargo blocks, etc., may be used as weights, and as they do not touch the deck at the lowered or inboard position, there is no fear of damage to them.

Heavy Lifting Derricks.—At the main hatch (usually No. 2), a "Jumbo" derrick designed to lift heavy weights will be found standing on end, and clamped to the foremast.

The weight it is capable of lifting safely will be stamped thereon, as with other derricks, and the average load will range from about 10 to 40 tons. However, with modern equipment such as the "Stülcken" mast and derrick, lifts of up to 300 tons are possible.

Deck Cranes.—Electrically-driven deck cranes have become popular and are often used in conjunction with the ordinary derricks. They have the advantage of being completely self-contained and are simple and easy to operate.

Topping Lifts.—The only topping lifts found on board are those rigged on derricks, sounding booms, log booms, light booms, boat booms and on the mainmast gaff. They usually take the form of a wire pendant, with a tackle attached for hoisting purposes, and it will be found that while the light and boat booms are generally fitted in this manner, the sounding boom, log boom and gaff are, as a rule, just single wires which hold the respective booms at a set angle.

In the case of derricks, three types are in general use, but neither one appears to be more popular than the other. They are as follows:

Tackle Topping Lifts.—The first, although not necessarily the best, is a tackle fitted between the head of the derrick and the upper table, with the hauling part led through a lead block on deck to the nearest winch.

Span and Tackle.—The second type has a heavy wire span reaching from the derrick head (when lowered), through a lead block on the upper table, and immediately below the block a tackle is attached, the other end of which is shackled to the deck. In this case the hauling part is led from the bottom block straight to the winch. Fig. 39.

Span and Chain Preventer.—The third rig is similar to the last type as far as the span is concerned, but instead of a tackle there is a single wire bull rope hanging to the end of the span, which is either taken direct to the winch or, in some cases, it will pass through a lead block first. Fig. 40.

When tackles are fitted, the weight of the derrick hangs on the tackle, but a bull rope is only a means of heaving the span down to

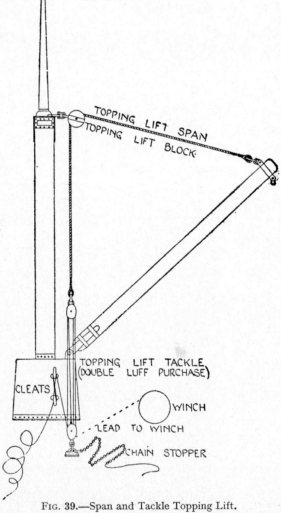

Fig. 39.—Span and Tackle Topping Lift.

the deck for the purpose of shackling on a chain preventer. After this is done, the derrick is lowered to a convenient height for working cargo, and the lower end of the preventer shackled to the deck. In this case the whole weight of the derrick hangs on the preventer.

Sending Up Topping Lifts.—Topping lift spans are seldom taken down at sea and stowed away like the rest of the derrick gear;

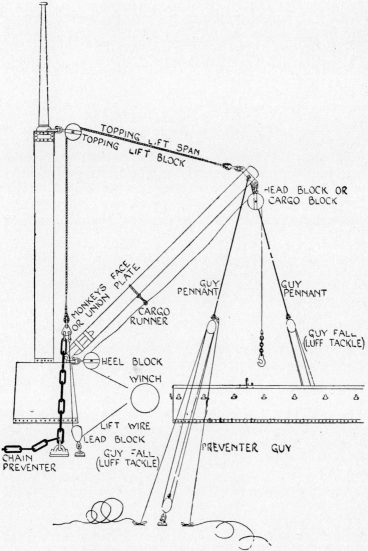

FIG. 40.—Topping Lift Span and Chain Preventer.

they are simply hove tight with a lanyard made fast to the nearest shroud or swifter. With a gantline, rove through a tail block made fast to the span itself, the tackle is hoisted in position for shackling on; but be sure to make the gantline fast around all parts, just below the block. Although it will be all right for hoisting like this in some cases, in others, according to circumstances, it will be better to seize the standing part of the gantline to the side of the block, to keep it upright.

Reeving Topping Lift Wires.— When rove off, the standing part of a topping lift wire is shackled to a lug on one of the blocks, but having a thimble spliced in this end it cannot be rove towards the lug in the same way as a plain end. Instead, the plain end of the wire is rove off the opposite way round, which leaves the standing part clear for shackling on.

With a heaving line married to the end of the wire, and rove through a sheave aloft, the parts are pulled through one at a time, but instead of climbing aloft each time to reeve the line, it can be bent to a part moving up to the top block, where it will be rove again by a man stationed there.

Oiling Topping-Lift Wires.—Marry an old gantline—preferably a topmast one—to the end of the hauling part of the wire, unshackle standing part from the block at head of derrick and lead it to a winch, through a snatch block if necessary. Make fast on centre drum and oil the wire as it is being hove on. When finished oiling, take the gantline to another winch and heave the topping lift back through the blocks again, finally shackling the standing part back in its place.

Shackling on Guys.—There is little difference in the size or length of guy pendants or falls, but in some cases the outboard pendant, on account of bearing a greater strain, is made of slightly heavier wire. Shackle pins should be secured with a mousing, and should be placed so that the top of the pin is uppermost when the derrick is in the air.

Preventer Guys.—When the union purchase system of working cargo is being used, the outboard guys on both derricks are fitted with a wire shackled to the pendant or passed over the derrick head. This is made fast on deck to act as a safety measure in case the guy should part.

Reeving Cargo Runners.—New cargo runners invariably have a thimble spliced in the end, and as this will not pass through any of the sheaves, the other end is rove instead. Starting with the head block, it is next rove through the lizard, or a roller (if any), then through the heel block to the winch, where it is made fast temporarily, as it cannot be run on the winch barrel until the derrick has been lifted.

Marrying Cargo Runners.—To reeve off a new cargo runner the end of the old one must be unrove from the heel block and the end of the new one married to it. By this means the new runner is easily pulled through the head block.

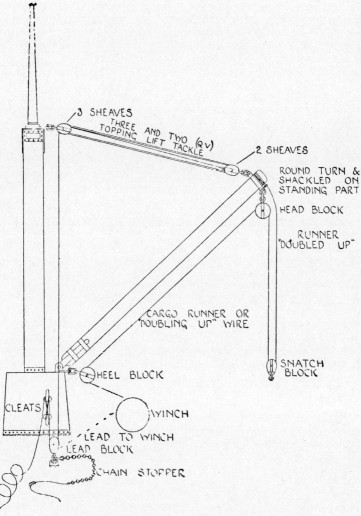

FIG. 41.—Tackle Topping Lift and " Doubling Up Gear."

To marry them firmly, take the bight of a few yarns and slip a marline-spike hitch over the end of one of the runners, beyond the serving, to prevent it slipping off. Lead both parts across the end of the wire and overhand knot them to keep both tightly in place.

Do the same with the second runner, then take the surplus ends from each runner and reef-knot them together.

The runners are now united and ready for reeving.

Doubling Up.—When a lift is considered rather heavy for an ordinary cargo runner to handle, it is "doubled up" to give greater lifting power. Actually, there is not enough length in the ordinary runner to do this, so an extra long "doubling-up wire", kept specially for the purpose, is rove instead. When a few fathoms have been pulled through the block, the end is passed around the head of the derrick and shackled to the standing part of the wire, but it must be done in such a manner that the turns will not slip down the derrick. To prevent chafe and scars on the paint, a thick layer of bagging is always placed under the wire. We now have a bight hanging from the derrick head, and on it a snatch block is hung with the hook hanging down, so that it may be attached to the slings on the lift. The hauling part leads to the winch in the ordinary way through the heel block. Fig. 41.

Rigging Heavy Lift Derricks.—The principal difficulty with heavy lift or "Jumbo" derricks, is that they have to be rigged while clamped to the foremast, in a vertical position.

All "Jumbos" appear to be rigged in much the same style. They have a topping lift tackle stretching from the derrick head to the foremast, and a purchase tackle hanging from the derrick head, the bottom end of which is attached to the heavy lift.

The three blocks which have to be sent aloft are hoisted with a wire gantline rove through a lead block on the foremast. Some vessels have a special eyebolt on which to shackle this lead block, otherwise a rope or wire strop must be used. The principal thing to watch when a strop is used, is that it will be high enough to allow plenty of drift for manoeuvring the blocks into position.

If the lead block is a heavy one, it will be necessary to rig a tail block and gantline, and this can also be used for sending up the end of the wire gantline.

After reeving off the topping lift and purchase falls, both hauling parts are led to the centre barrel of the winches, either through heel blocks in the case of small vessels, or special lead blocks bolted to the deck in big ships. This enables both the derrick and lift to be hove up or lowered down at any moment while hoisting the load on board. In some cases the purchase hauling part leads from the bottom block, through a lead block near the derrick head, and then down to the special lead thence to the winch.

Owing to its vertical position, the derrick must be lowered until the topping lift takes the weight, and this is done with a wire led from the forecastle head to the top of the derrick. Even when the clamp and lashings are removed, it may be necessary to pull down on one of the guys to start the derrick lowering away. Fig. 42.

Sending Up Derricks.—The first thing to do is to make sure the lashing on the derrick head and crutch is cast adrift. This is very

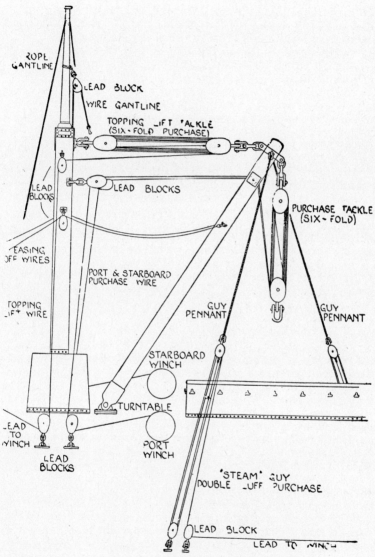

ROPE GANTLINE

LEAD BLOCK

WIRE GANTLINE

TOPPING LIFT TACKLE
(SIX-FOLD PURCHASE)

LEAD BLOCKS

LEAD BLOCKS

PURCHASE TACKLE
(SIX-FOLD)

EASING OFF WIRES

PORT & STARBOARD
PURCHASE WIRE

GUY PENNANT

GUY PENNANT

TOPPING LIFT WIRE

STARBOARD WINCH

TURNTABLE

LEAD TO WINCH

LEAD BLOCKS

PORT WINCH

'STEAM' GUY
DOUBLE LUFF PURCHASE

LEAD BLOCK

LEAD TO WINCH

FIG. 42.—Heavy Lift Derrick.

often forgotten, and minor damage usually results from the omission.

It is not always necessary to spread both guys, unless the vessel is at sea and rolling slightly, or the derrick has to be kept clear of the front of the bridge, whistle lanyards, etc. Very often in port, if on an even keel, the guys are spread after the derrick is on end, and this saves much unnecessary labour as well as time. As a rule, the weather guy, or one on the high side (in the case of a listed vessel), is sufficient for steadying purposes, but everything depends on circumstances.

When a vessel is rolling more than slightly, it is a good plan to pass a rope over the derrick end, make one part fast to a ring-bolt on the coamings, and pay out on the other part as the derrick goes up. This will prevent any violent swinging and take the weight off the guys, and as only a bight is over the derrick end the rope is easily unrove again.

If wire preventers are to be used, they must be shackled to the pendants before heaving the derrick on end.

When the hands at the guys are ready, take the topping lift fall to the winch and heave away, but make sure the end of the cargo runner is fast or it will go up with the derrick. When the runner plumbs the section of the hatch which is to be worked, the topping lift fall is stoppered off, made fast on the crosshead or cleats, and the guys spread in their proper positions.

Should the topping lift be the single span type, which hangs to a chain preventer on deck, the end of the span will have to be hove down to the deck to make the connection, after which the derrick is lowered to the required position, and the preventer is shackled to the lug on deck at the nearest link.

Lowering Derricks.—If derricks are to be lowered down ready for sea, and the gear is to be stripped and stowed away, it is easier to unreeve the runners before lowering, unless the vessel is rolling, in which case the slack of the runner is hove in as the derrick comes down, to prevent it swinging about.

Be sure to lower the proper one first, otherwise, in the case of derricks which cross, they will not fit in the crutches properly.

The procedure as regards actual lowering is different in each ship. In some cases the fall is slacked away on the crosshead or cleats where it is made fast, while some are "stoppered off", transferred to a winch end, or run on the centre barrel and lowered by working the steam valve and reverse lever.

If a single span and chain preventer is fitted, the derrick will have to be hove up first, so that the span will come down to the deck for unshackling the preventer. When about to drop into the crutch, it is usually necessary to pull the head block and runner clear, and the man doing so should immediately pass the lashing if it is to be put on.

Once the derrick is landed, it will be safe to take the fall off the winch, and if the topping lift is a tackle shackled to the deck near the bulwarks, it will probably be lashed up and down to one of the shrouds, but if placed near the mast, the hauling part is wrapped around its own parts, and the lot is bowsed to a nearby cleat or pin.

When rope falls are in use a certain amount of slack should be allowed for shrinkage in wet weather, otherwise the derrick might lift out of the crutch if this is not done.

If a derrick is not to be stripped the runner is hooked to the heel block, or made fast with a good strand of yarns and hove taut on the winch. The guys are also shackled on in the same place, hove taut by hand, made fast, and the ends coiled over all parts.

Lashing Derricks.—No. 1 hatch cargo derricks are situated right forward, so they are naturally exposed to any seas breaking over the forecastle head. For safety, extra lashings are passed around them; these are usually rove through ringbolts on the coamings, a round turn is taken around each derrick, and the lashing is worked from one side of the hatch to the other as many times as might be required. All parts are frapped underneath the derricks to make them more secure, and old bagging or burlap is placed in between the lashing and the tarpaulins to prevent any chafe.

Frapping.—This is a method of tightening up a lashing. The spare end (or sometimes a piece of small stuff) is passed around two or more parts and hauled tight, especially where there is a long drift. Frapping doubles the efficiency of a lashing. This is even more so in the case of barrels. If the lashings are in any way slack barrels very soon work loose and get adrift.

Driving Winches.—Any person unacquainted with the operation of a winch—either steam or electric—should never attempt to take over the controls until they have received instructions as to the proper way of handling them. Carelessness in this direction has resulted in many nasty accidents.

Complete details concerning the driving of all types of winches cannot be given here, nor would they be much use without practical demonstration.

CHAPTER V.

TANKERS AND OTHER DEVELOPMENTS

General Tanker Knowledge and Procedure. Refrigerated Ships. Modern trends in other Cargo Equipment. Masts and Derricks. Cranes. Steel Hatch Covers. Containerisation. Unitisation. Palletisation. Roll-on, Roll-off. Future Trends.

GENERAL TANKER KNOWLEDGE.

Tankers.—More than one-third of the world's merchant vessel tonnage consists of tankers for the carriage of cargo in bulk which,

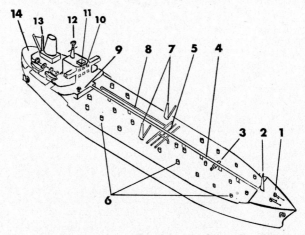

Fig. 1.

1. Forecastle with capstans and cable lifters.
2. Foremast, carrying navigation light and outlets for gas lines.
3. Mooring winch.
4. Flying bridge.
5. Pipe manifold.
6. Tank lids.
7. Midship samson-posts with 5-ton derricks to handle pipelines.
8. Overdeck pipelines.
9. Pumproom.
10. Navigation bridge.
11. Compass platform.
12. Mainmast fitted with navigation and signal lights in addition to radar scanner and radio aerials.
13. After samson post with derrick for handling stores.
14. Docking bridge.

for the most part, is concerned with oil in its various stages of refinement. They vary in size from small coastal tankers to the largest vessels that have yet been built. The many different types of liquid cargoes need special designs, but, regardless of these points, all tankers have certain similarities.

All tankers have the engines placed aft to avoid the difficulty of making the shaft tunnel oil-tight. They were mostly of the three-island type with forecastle, poop and amidships navigation bridge space, but many tankers now have the bridge placed on the fore-end of the after housing to leave the main deck clear for cargo working. (see sketch, p. 130). The main deck cargo pipelines, gas-lines and relatively small tank lids, indicate the kind of cargo carried.

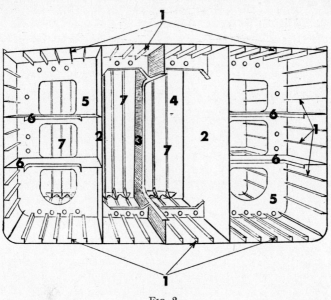

Fig. 2.

The above sketch shows the main features of a cross-section of a tanker.

Longitudinal frames (1) running lengthways along the under side of the deck plating and the inner side of the shell and bottom plating, together with the two longitudinal bulkheads (2) and a heavy centre-line web structure (3), all contribute to the longitudinal strength.

Strength across the hull is provided by the transverse bulkheads (4) and the widely spaced transverse webs (5). Stringer plates (6) strengthen the wing tanks and vertical stiffeners (7) are shown on the transverse bulkhead. Both these systems must be secured to each other to effectively spread the load at any point.

This construction is particularly suited to the type of liquid or bulk cargo carried by tankers but would not be suitable for dry-cargo ships because the deep transverse webs would cause a considerable amount of broken stowage. However, there is an alternative system of longitudinal framing along the deck and bottom plating combined with transverse framing down the shell plating at the sides which is used in both tanker and dry-cargo construction and some later vessels have been constructed to serve as either oil, bulk or ore carriers (O.B.O.).

Cargo Tanks.—Two oil-tight fore and aft bulkheads extending the full length of the ship's cargo carrying space form a large centre tank with smaller port and starboard tanks at the side. Many bulkheads placed athwartships, for the full width of the ship, form sets of tanks. The average tanker has nine or eleven sets, making a total of 27 or 33 tanks. In large crude oil-tankers, some wing tanks are not needed for cargo, in which case these tanks are isolated and provided with their own pumping system to be used solely for water ballast.

PIPELINE SYSTEMS.

There are many variations of three main systems. For clarity, only the pipelines are shown in the following diagrams, but it should be remembered that each tank in both the Ring Main and the Direct Pipeline systems will have a tank suction connected to the pipeline.

Ring Main.—This consists of pipeline loops from the pump-rooms placed within the range of the tanks. Crossover lines connect the port and starboard sides of the Main Ring at each set of tanks. By opening valves in these crossover lines, the cargo ring circuits

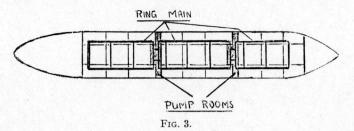

FIG. 3.

may be reduced to encompass a lesser number of tank sets. With the two pump-rooms forming part of cofferdams, this system lends itself to the easy separation of parcel cargoes on general purpose tankers.

Direct Pipeline.—Pump-rooms, if situated within the range of cargo tanks, include, for purposes of efficiency and safety, reciprocating pumps for discharging cargo. Centrifugal pumps have a much higher discharge rate than reciprocating pumps, but require powerful rotating machinery to drive them. For this reason, tankers using such pumps have them situated at the extreme after end of the cargo-carrying compartments. A driving-shaft is then positioned to couple the pumps to their associated drive motors situated at the forward end of the engine-room, the rotating drive passing through bulkheads with an oil-tight seal.

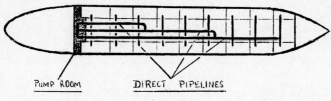

PUMP ROOM   DIRECT PIPELINES

FIG. 4.

Freeflow System.—This system is used by large crude oil tankers engaged in the carriage of one class of oil at a time. The pipeline is dispensed with, and valves are set in the bulkheads to allow a free flow of oil from forward to aft to the main suctions in the after centre tank.

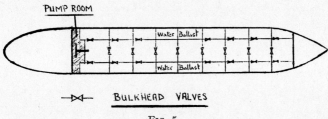

PUMP ROOM

Water Ballast

Water Ballast

—▷◁— BULKHEAD VALVES

FIG. 5.

Stripping Lines.—Centrifugal pumps require to be primed by a sufficient head of oil in the tank to be discharged, otherwise the pump impeller will accelerate, and this may lead to serious damage. Such pumps are not suitable for discharging the last foot of oil in the tank, so that a separate stripping pipeline of smaller diameter is fitted, and the tanks are then drained by a small reciprocating pump. More recently, centrifugal pumps have been fitted with special priming devices so that they may be used to drain the tanks.

Deck Lines.—The cargo lines, already described, pass along the tank bottom to the pump-room. Pump pressure then causes the liquid cargo to pass upwards to other cargo lines at the main deck level, which convey the cargo over the deck to the amidships manifold to which flexible shore lines are connected.

Valves.—Oil flowing through the main lines situated above the tank bottom is controlled by sluice valves. These valves are operated from the main deck level by spindle extensions. The upper part of the spindle projects a reasonable distance up through the main deck by the use of oil-tight glands, and a hand wheel is attached.

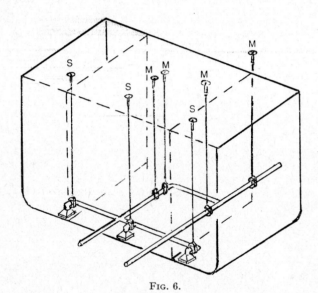

Fig. 6.

The suction valves (S) control the flow to and from the tank. Master valves (M) direct the flow along the main lines. In the ring main system, the flow across the vessel depends on crossover valves. To enable sea-water to be used as water ballast, sea valves are fitted in the pump-room between the sea and the pipeline. Other sluice valves are fitted in the deck lines and the manifold.

Colour Codes.—To identify the valves, the valve wheels are painted to conform to a colour code, which may vary from company to company. As an example, the following code has been used:

Port tank suction valve	Red
Starboard tank suction valve	Green
Centre tank suction valve	Black
Master valves	White
Sea valves	Blue

Cofferdams.—These spaces form a full-width separation between tank systems or between cargo tanks and the engine-room. Pump-rooms are built into suitable cofferdams.

Pump-rooms.—In addition to the cargo pumps, the pump-room contains stripping bilge pumps as well as a complete pipe and valve system to transfer the cargo to the main deck level. A pump-room is fitted in the forward cofferdam, which has a dual purpose. Firstly, a pump is fitted to transfer fuel oil from the forward deep tanks and secondly, a separate pump is used to ballast the Fore Peak and to pump the bilges. This pump may also be used as an emergency fire pump.

Fig. 7.

Tank Lids.—Tank lids stand about three feet high and provide a protected access to the cargo tank. There are many different types, but all of them have the features shown in the above sketch. The lid of the hatch is made fluid and vapour-proof by a channel of greasy packing, which is pressed firmly on to the coaming of the hatch when the lid is securely battened down by the use of attached wing nuts. A sighting port is set into the lid. Various means of raising the lid are in use, but the worm-screw method illustrated in the sketch is commonly employed.

Gas-Lines.—Vapour released from the surface of oil within any tank is vented to the atmosphere via small pipes about six inches in diameter called gas-lines. These couple each tank hatch coaming to a larger gas-line, which passes up the mast. At a safe height, this larger gas-line terminates in a spark-proof ventilator. Individual valves are fitted to separate various sections of these gas-lines. In addition, pressure-vacuum valves are fitted to provide automatic control of pressure inside the tank to within pre-set limits.

Flying Bridge.—This structure provides safe passage for the crew between the after sections of the vessel to the amidship superstructure, and then forward from the amidship structure to the forecastle. A tanker when fully loaded in a seaway frequently ships heavy seas on the maindeck, because she is permitted to load to a deeper draft than an ordinary dry-cargo vessel of similar size. The provision of safety for the crew is thus essential. The underneath of the raised flying bridge conveniently provides a safe position for wash deck lines, steam pipes, compressed air lines, electrical cables and similar services, thus avoiding a clutter of pipes on the main deck.

Oil Measurement.—To measure the weight of oil, it is necessary to know the volume and density of the oil. It is then a simple case of multiplication:

Weight of oil = Volume of oil × Density of oil.

The volume of each tank is calibrated by the shipbuilder for every 3 or 6 in. of depth of tank, and so, to obtain the volume of oil, it is only necessary to measure the depth of the oil. This can be done by lowering a weighted tape to the bottom of the tank to give a

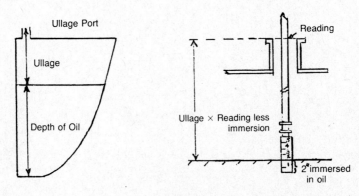

FIG. 8.

direct measurement or, alternatively, the ullage or height of the space above the oil can be found by lowering a measuring tape or ullage stick to the surface of the oil. The ullage is measured below a fixed point, which is usually the top of the ullage port. The ullage or depth of oil is converted to volume by using the tables supplied by the shipbuilder.

More sophisticated methods of oil measurement are now in use. In one system, the ullage is measured by reading a steel tape in a gauge on deck. The tape is spring-loaded and extends to a float on the surface of the oil. A newer system uses pressure gauges to give both the ullage and a direct reading of the weight of the liquid. These new systems, together with the development of compressed-air actuators which open and close valves, have led to centralised control systems, which, in turn, have increased the efficiency in all tank operations, and effectively reduced the amount of labour involved.

Density.—To find the density of the oil, a cargo sample from any particular cargo is obtained by using a "dip can". A hydrometer floated in the sample enables the specific gravity of the sample to be obtained. The specific gravity of the cargo is the number of times the oil is heavier than an equal volume of pure water. Oil being lighter than water, the reading will always be a decimal, and the density of the oil is equal to the density of pure water multiplied by the specific gravity of the oil. Changes of temperature affect the density and a standard temperature of 60 degrees Fahrenheit has been fixed by the Oil Industry. Tables have been published and are always available, so that the specific gravity of the sample can be adjusted or corrected for temperature differences observed between that of the sample and the standard in use.

The oil hydrometer is made of glass and the instrument is weighted to float upright in liquids of a particular range of specific gravity, which is indicated on a scale on the narrow, hollow stem forming the upper part of the instrument. Sets of hydrometers are supplied to cover the range to be carried in the ship.

Loading.—Before loading, the amount of cargo required to load the vessel down to the load-line is estimated, so that a loading plan can be prepared to ensure that the vessel, when loaded, should be on an even keel without overstressing the hull.

After discharging any ballast from the cargo tanks, the sea-valves are closed and sealed. All tanks are inspected by a representative of the cargo owner or shipper, and when the inspector is satisfied that the ship's tanks are properly drained and otherwise in a reasonable condition for the carriage of the cargo specified for loading, the tank lids are secured and the gas-line set to allow the gas to flow out through the mast line. The cargo valves are adjusted so that the oil may flow to the tanks required. Flexible shore pipelines are connected to the main deck manifold and, when the ship is ready to commence loading, the shore and manifold valves are opened. At first, the flow is restricted. Tanks at each end of the tank system are opened and, when it is clear that the lines have been opened correctly, other tanks are opened and the loading rate increased to the maximum permitted. Open ullage ports are covered with wire gauze.

The order in which the tanks are loaded will depend on the usual practice of the company and will take into account of the type of cargo, the rate of loading and the possible stresses on the hull. One tank amidships, usually No. 6 centre, will be the last tank to be loaded, and the loading rate is reduced so that loading can be stopped as soon as the load-line shows that the vessel is fully loaded.

Since the shore-line may be cleared of oil by blowing through the line with compressed air, the tank valve must not be closed until the shore staff indicate that it is safe to do so.

Oil measurement commences as soon as a tank is filled. Ullages, temperatures and specific gravity are obtained. The water content is estimated by lowering a rod covered with water-finding paste to the bottom of the tank and measuring the length of discoloured paste. This dip is converted into water-tons and subtracted from the oil tonnage.

When the oil measurement is completed, the shore-line is removed and all valves closed. Ullage ports are closed and secured. Pressure/vacuum valves within the gas-lines, which were lifted before loading, are now set to the working position. The draught is read and, with the paper-work completed, the vessel is ready to sail.

Cargo Heating.—Many oil cargoes in a less-refined state are viscous at ordinary temperatures. Some heavier grades cannot be pumped unless their temperature is raised. Precise instructions are issued by the oil company about any cargo-heating, and these must be carefully followed. Heat is supplied by steam passing through metal coils fitted to the bottom of the tank. The temperature of the oil must be taken at regular intervals and the steam valves adjusted accordingly.

Discharging.—Whilst the shore-lines are being connected and the ship's pipelines set for discharge, all the ullages, temperatures and water-dips are taken. Later, the amount of cargo on board on arrival will be checked against the total loaded.

Ullage ports are opened to prevent any partial vacuum in the tank, and the opening covered with wire gauze. Heating coils are shut off either at the commencement of the discharge of the tank or just before the level of the cargo is brought down to the heating coils.

The order of discharge is carefully planned to give the fastest and most efficient discharge. In most cases, the final draining of the tanks has a major influence on the order of discharge. Each tank is discharged by the main cargo pumps to the lowest possible level, and the drainings are stripped into one tank called the slop tank. Ideally, only a short period of time should separate the completion of the discharge by the main pumps, the completion of stripping the tanks and the discharge of the slop tank.

As soon as the cargo is discharged, each tank is dipped to find the amount of cargo which it has been impossible to pump out of the tank. This residual cargo may be due to the oil level having fallen below the level of the suction strum box, limber holes in the fore-and-aft or athwartship girder work being blocked, or the amount of oil in the tank having formed into a sludge. A shore representative will check to ensure that each tank is as empty as possible. The shore-line may then be removed and the pipelines set for ballasting.

Ballasting.—Tankers are engaged in the carriage of cargo for about a half of their operating time, which means that water ballast must be carried for the remaining period when the vessel is proceeding from a discharge port to the next loading terminal. Since the water ballast is taken aboard at the discharge port, the cargo tanks used for ballast purposes will still be contaminated when on voyage to the loading port. The ballast aboard is changed during passage by putting clean sea-water into other tanks which have been cleaned and discharging the initial ballast into the sea. Some larger ships have several tanks reserved solely for the carriage of ballast. Such permanent ballast tanks have the advantage of reducing the amount of cleaning required. The amount of ballast used, together with its distribution, should be sufficient to make the vessel manageable in a seaway without causing unnecessary hull stresses.

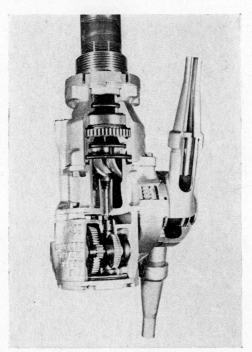

Fig. 9.—Butterworth Tank-cleaning Apparatus.
(With section of casing removed.)

The ballast entering cargo tanks at the discharge port becomes "dirty" ballast by contamination with the remains of the previous cargo. When all the tanks without ballast have been cleaned and the pipelines washed through, "clean" ballast is introduced into some of the washed tanks. With this operation complete, the original ballast is pumped overboard and the tank cleaning resumed. This process is not so simple as it may appear, because not only must the vessel be suitably ballasted at all times, but it is an offence to discharge oil-contaminated water into many areas of the open oceans, in addition to all coastlines and neighbouring waters.

Tank Cleaning.—On the outward passage to the loading port, the important operation of tank cleaning must be carried out. On the main deck above each individual cargo tank, circular plates are bolted down. Removal of the nuts, together with the plate, reveals an opening about 15 in. in diameter, through which patent tank-cleaning apparatus may be introduced. The "Butterworth" tank-cleaning apparatus shown in Figures 9 and 10 is extremely popular.

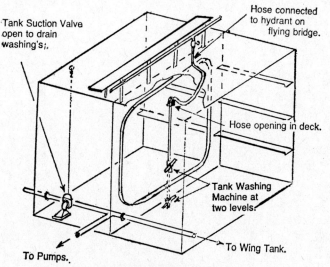

Tank Cleaning Arrangements
Fig. 10.

Simple to use and easy to maintain, such appliances reduce the number of man-hours which would otherwise be required to satisfactorily clean tanks. The tank-washing apparatus is attached to the end of a stout rubber hose and lowered into the tank. The weight of the complete mechanism is supported on a crutch, mounted concentric to the circular deck holes. Water at high pressure causes the machines to rotate horizontally, whilst the jet ends rotate at the same time in a vertical plane. Thus, powerful jets of water are directed to all parts of the tank. To make sure that this is carried out effectively, it is usual to wash at three different levels for a set period of from 20 to 90 minutes.

The temperature of the water used for tank cleaning will vary with the type of cargo previously carried, and though, in general, most of the washing uses cold water, for a hot water wash, particularly after carrying crude oil, the temperature of the water must be maintained at about 150 to 180 degrees Fahrenheit.

When the oil has been washed from the tank, the various pipelines and pumps must also be washed through by a somewhat complicated procedure, which ensures that every section of line is cleaned.

Crude Oil Washing.—Using the ship's cargo as a washing medium through fixed washing machines has considerable advantages. Washing commences during discharge and most of the deposits, sludges and waxes are sent ashore with the cargo. The tanks are kept full of inert gas throughout the operation.

Gas Freeing.—The tank-washing will not have cleared all the oil from the compartment. If hot water has been used for cleaning purposes, then oil will have been made to vapourise, in addition to the normal vapourisation caused by agitation from the washing-machine jets. The older method of setting up windsails to clear tank gas was effective, but has recently been considered too slow and cumbersome. A quicker method is to use portable fans driven by compressed air or water. If used in pairs per tank, the speed of "gas freeing" is greatly increased. When used in this way, one machine introduces clean atmospheric air to the tank top, whilst the other extracts air from the tank bottom by means of a long pipe extending the full depth of the tank. An explosimeter is used to check the gas content by drawing a sample by means of bellows through a tube and thence to the meter, which will immediately register the presence of gas. This meter should always be used before any person is allowed to enter a tank.

Once the tanks have been made gas-free, a working-party can finish off the final stages of cleaning by removing any sludge or rust scale, which can be put into buckets and hoisted to the deck. The gas level must be checked at regular intervals whilst this work is in progress.

SAFETY.

The dangers involved in the carriage of oil cargoes are well known. Tanker owners and oil installation operators issue detailed safety precautions to cover all foreseeable conditions and contingencies. The following indicates some of the main requirements:

During Cargo Handling.—Strict compliance is necessary with any rules concerning smoking on board or the use of the galley fires, or of any ship's electrical appliance. Fire appliances should be placed ready at hand, e.g. hoses with spray nozzles should be rigged in the vicinity of the main deck manifold, supplemented by two-gallon foam extinguishers placed on the flying bridge.

Oil spillage may be a potential source of danger, and should be dealt with as promptly as possible by having suitable equipment ready at hand.

Accommodation openings should be kept closed or at least covered with sparkproof gauze, and similar gauze shields should be placed over any open ullage or sighting ports during cargo working. All scuppers on the main-deck should be plugged to reduce risk of oil entering the harbour area.

A potential source of danger exists if the tanker is at a different electrical potential to that of the oil jetty, which could cause an electrical discharge leading to an explosion. To avoid this, a bonding wire from the jetty is clamped on to the steelwork of the sheerstrake to provide an earth.

Finally, to cover the case of an emergency occurring which would require the tanker to be towed from the berth, tow wires are rigged both forward and aft at the mooring stations on the offside. The outboard ends of these tow wires have an eye splice and the inboard ends are turned up on the bitts. The towing wires when positioned are of adequate length so as to nearly reach the waterline.

During Tank Cleaning.—This is a period of some danger with large quantities of gas being driven from the tanks and the mixture of gas and air in the tanks at its most explosive level. A ban on smoking and other precautions may be necessary even in the accommodation. The tank-cleaning party, in particular, must avoid causing a spark. Cigarette lighters and any similar items should not be carried whilst working on deck. The tools used on tank-cleaning equipment should be made of a material such as bronze, which will not cause a spark if dropped on to the steelwork of the ship. For this reason, steel tools should never be used under such conditions.

To prevent a static electricity charge, which may have been produced by the water in the hose, causing a dangerous discharge in the tank, the tank-cleaning machine is always earthed to the deck by means of a bonding wire. A safety wire is attached to the machine in case it becomes disconnected from the hose.

Petroleum Gas.—Petroleum gas is toxic. No man should enter any compartment which contains gas unless he is wearing a self-contained breathing apparatus. The presence of gas can be ascertained by an explosimeter, but in the absence of such a meter, it would be safer to take full precautions. In the event of a member of the crew becoming overcome by gas, it is essential that his rescuer should wear a suitable breathing apparatus. The victim should be removed to a position free from gas and artificial resuscitation, amyl nitrate, ammonia capsules or oxygen apparatus applied.

Inert Gas.—A fire or explosion needs (1) fuel, (2) sufficient oxygen, and (3) a source of ignition. Usually there is plenty of fuel on board a tanker. Most safety regulations, such as restrictions on smoking, are aimed at preventing ignition being applied to a mixture of fuel and oxygen. Inert gas systems reduce the oxygen content of the mixture to a safe level, where there is insufficient oxygen to support combustion.

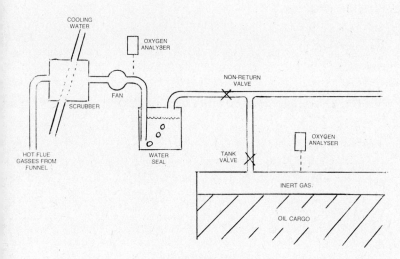

Fig. 11.—Inert Gas System.

In most inert gas systems, the hot flue gases from the boilers are passed through a scrubber. Here the gases are cooled and cleaned of corrosive products. The cool, clean, inert gas is delivered by means of a fan to the tanks. To prevent gases flowing back along the gas-lines from the tanks to the machinery space, a water-seal and non-return valve (N.R.V.) is fitted. The oxygen level of the inert gas is checked both in the supply line (*see* Fig. 11) and by taking readings in each tank. If the oxygen content has risen above a safe level then the inert gas content is increased until the vapour in the tank is again safe.

General Safety.—Besides the special equipment required for the oil trade, tankers carry the usual equipment required to be carried by other types of vessels.

TERMS USED.

Black Oils.—Black oils are those which are roughly classified by colour and include crude oils, fuel oils, certain diesel and gas oils.

Clean Oils.—Clean oils are those which are almost colourless due to refining, such as lubricating oils, petrol and kerosene.

Crude Oils.—Crude oils are the natural product obtained from the oil-wells. The characteristics of crude oil vary greatly between the different fields.

Flash-Point.—Flash-point is the lowest temperature at which the oil will give off sufficient vapour to produce an explosive mixture when mixed with a sufficient quantity of oxygen.

CLASSIFICATION OF CARGOES.

Class A.—Oils with a flash-point below 73° F. (Crude oils, motor and aviation spirits, etc.)

Class B.—Oils with a flash-point between 73° to 150° F. (Kerosene, white spirit and some crude oils.)

Class C.—Oils with a flash-point above 150° F. (Gas and diesel oils, fuel oils and lubricating oils.)

REFRIGERATED SHIPS.

Refrigeration.—Vessels fitted for the carriage of refrigerated cargo, i.e. frozen or chilled cargo such as meat, or where temperature control is essential as in the case of the carriage of fruit, are usually equipped with a vapour-compression system of refrigeration using carbon dioxide gas or freon as the refrigerant. The gas passes through a compressor where it is compressed, thence to the condenser where it is cooled by sea-water and liquified and thence by an expansion valve (the opening or closing of which controls the temperature) to the evaporator where it evaporates by extracting its latent heat of vaporization from brine in surrounding brine pipes. In this way the brine may be cooled to any temperature down to 0° F. and becomes the refrigerant for cooling the cargo spaces by the "Grid" or "Air Battery" systems.

Calcium Chloride.—A solution of calcium chloride and water in the ratio of $3\frac{1}{2}$ lbs. of chloride to each gallon of fresh water is the basis of the brine referred to above and has a freezing-point of −21·5° F. Other chemicals are added as a protection against corrosion troubles and lime is used to neutralize any tendency to acidity.

Grid System.—In this system the cargo spaces are lined with $1\frac{1}{2}$-inch diameter pipes at about 4 inch centres and divided into a number of independent circuits. The cold brine as the refrigerant, is pumped through these pipes to cool the cargo spaces and any heat which may be absorbed by the brine, is extracted in the evaporator as the brine circulates round.

In the final stages of loading, grids of piping are connected up in the square of the hatch to complete the circuits and as the produce is already frozen or chilled down to temperature, the hatches are then sealed and insulated to avoid the ingress of heat. There is no ventilation or circulation of air.

Air Battery System.—This system of refrigeration is the most efficient for the carriage of fruit which being a "live" cargo gives off carbon dioxide gas and tends to raise the temperature. This is counteracted by the air in the hold being brought to a controlled temperature and circulated through the cargo by means of fans.

The air is first cooled by passing over a closely spaced bank of brine pipes and then fed through a system of delivery and suction trunks by the fans to ensure uniform distribution. The amount of carbon dioxide will increase and is measured by special Carbon Dioxide Indicators. The concentration should not be allowed to rise above 2% before ventilation is started to change some of the air.

This system can be used equally as well for frozen cargo usually carried at a hold temperature of 14° to 16° F. or for chilled cargo usually carried at about 28·5° F. and for this reason is being fitted in most new constructions.

Temperatures.—In earlier days a considerable amount of time was taken up in checking temperatures which was a manual process whereas today all refrigeration systems incorporate distant reading thermometers of an electrical or a liquid-expansion type. The leads are now brought to a central point which is usually in the compressor room so that information is immediately available for all parts of the vessel. This also applies to the Carbon Dioxide Indicators used in the Air Battery system.

Insulation.—For efficient operation it is essential to avoid as much as possible the ingress of heat from outside to the cooled cargo spaces and this can only be carried out by the use of good insulating materials. The most common materials aboard ship are granulated cork and what is known as glass wool. At the sides of the ship there is a boarded up space of up to 12 inches or so which is filled with granulated cork and the same applies to the decks. The plug hatches are usually filled with glass wool and in a non-permanent way bags of sawdust may be used to complete the insulation in the hatchways.

Meat Hooks.—A similar type to those used in a butcher's shop are employed for the same purpose on vessels carrying chilled meat.

Meat Chains.—Frozen meat carcases being frozen solid are stowed solid but chilled meat is always hung. For this purpose hooks and chains are used and the chains are known as "single" or "double" according to their length.

Bull Bars.—For the carriage of chilled meat, galvanised iron bars are fitted in between the hatch beams and on these the carcases are hung to complete what would otherwise be wasted space in the hatchways.

Plug or Insulated Hatch Covers.—Refrigerated cargo space is insulated throughout and plug hatches are fitted under the ordinary hatch covers to insulate the space in the hatchways when loading is finished.

MODERN TRENDS IN CARGO EQUIPMENT.

In the days of sail, the masts were the means of transmitting the motive-power of the wind to the hull of the vessel and consequently safety demanded a complicated system of standing rigging to support the masts against all possible stresses. Today the much more powerful driving power is contained within the hull, leaving the masts and deck fittings in general for the sole purpose of the quick loading and discharging of cargo.

As a result it became the practice to have a strong lower-mast supported by shrouds and a stay, with a lower-mast table to take the heels of the derricks and a second outreach or table extending each side of the lower-mast head to take the blocks of the topping-lifts supporting the heads of the derricks. This upper table is still often called the "crosstrees" but whereas in the past the purpose of the crosstrees was to spread the rigging of the masts above, the sole purpose now is as an "outreach" for the derrick gear. In addition there was a short topmast (usually telescopic) supported by a light topmast stay and two backstays, one each side.

In port, when working cargo, the spaces on the deck adjacent to the hatches were cluttered with the portable beams and the wooden hatch covers whilst the standing rigging generally interfered with being able to swing the cargo across the deck. "Scotchmen" were fitted to take the chafe on the stays but it was often a case of time taken up in letting go any of the rigging which was in the way and the setting up of additional preventer stays.

Over the years the number of shrouds at the side of the masts have been reduced and ships have been operating with a minimum of standing rigging but the latest trend is for a mast structure strong enough to stand up to all requirements without the need for any supporting rigging. This is well carried out by the various "bipod" mast structures which not only give great strength but also provide means for ventilation of holds and are suitable for use with the various heavy-weight and special derrick systems.

Furthermore, the advent of the self-stowing steel hatch covers and the raising of the winches to a special winch deck has cleared the decks and added greatly to the speed and efficiency in working cargo.

The Crenn Mast.—This is a monopod which provides a shroudless mast unit. It is strongly welded to the main-deck usually at a point where the transverse and longitudinal bulkheads intersect. It includes a lower outrigger supported by lateral air intakes and the outrigger also conveniently takes the heels of normal derricks. It is built to take a heavy-lift derrick on the after side in addition to the usual four ordinary derricks.

Fig. 12.—Stülcken Mast.

The Stülcken Mast.—This is essentially a bipod mast specially intended for ships carrying heavy and very heavy weights. The derrick which is incorporated with the mast structure has its heel centrally between the two parts of the mast structure so that it can just as easily serve a hatch either forward or abaft its own structure without having to shift the heel of the derrick. There being no shrouds, the derrick can handle weights of up to 300 tons at a time on one side or the other without any adjustment of the gear. The structure can also accommodate the ordinary derricks without interfering with the heavy derrick which is kept topped up when not in use. The controller platforms or cabs are built in with the masts and are high enough to give a clear view for all working operations.

Fig. 13.—Hallen Universal Swinging Derrick.

The Hallen Swinging Derrick.—This equipment can be used with most mast structures of the pole or bipod type fitted with strong outreaches to take the two topping lifts by means of which the derrick is topped and swung. To ensure stability when slewing outboard, a heavy fore-and-aft "D" frame is fitted so that when the derrick is outboard, the offside inner guy pennant is in contact with the face of the "D" frame to maintain a working angle. If it is not desired to fit these heavy frames, the "Universal" type Hallen achieves the same stability by the alternative method of maintaining a controlling angle between the guys by the introduction of two universal swivel outrigger rods which are attached to the yardarm of the outreaches.

The two hauling parts of the two topping/slewing guys are led from the derrick head through lead blocks on the mast trunk to the barrels of two winches which are simultaneously controlled by one "joystick" type lever. This one lever completely controls the raising or lowering of the derrick head and at the same time the slewing to the right or left and a third winch with remote lever control operates the hoisting and lowering of the load. As in the case of a crane, one man has complete control and because of the great flexibility is able to spot-place the cargo on both ship and quay which makes it very suitable for dealing with containers or pallets.

Deck Cranes.—Hydraulic or electrically driven cranes are tending to replace some of the derricks and it is becoming increasingly common to find a hatch served by just one of the modern types of derrick together with one such crane.

The jib is usually of the level-luffing type which allows the jib to be raised or lowered but ensures the weight remains at the same level until hoisted or lowered by the normal control. These cranes may have a large reach and angle of slew and be fixed centrally or may be set on rails so that they can be positioned at one corner or the other of the hatch whilst working. They are controlled from a cabin which gives a clear view of the working area and are simple and easy to operate.

Containers.—Containers have been used to a limited extent for many years but it is only quite recently that the container, as it is now known, has come into real prominence and has created a completely new concept in the transport of cargo.

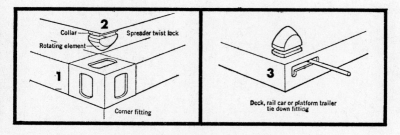

14.—Fig. 1. Container corner fitting.
            2. Conical dog twist lock device.
            3. Tie down fitting and twist lock.

The main advantage in using containers is the reduction in handling the individual cargo whether it is in boxes, cartons or crates. The ideal concept is that in which a manufacturer will fully load a container at his own factory and it will then be shipped to the purchaser of his product who will open and empty it. During the whole of the transhipment time the cargo within the container will not be subject to any individual handling thereby reducing the risk of damage or pilferage. Part loads are now being handled at inland assembly areas and full containers made up for eventual shipping.

The value of a container is obviously dependent upon the ability to move it from the place of origin to its destination without it being opened and this has produced problems of being able to handle such large and heavy loads.

The original container was built to be 20 ft. × 8 ft. × 8 ft., but under metrication this is now built to metric measure and is standardized at 6·1m × 2·4m × 2·4m though it may still be referred to as the 20-footer. In the same way, the largest container has been standardized at 12·2m × 2·4m × 2·6m, but may still be referred to as the 40-footer.

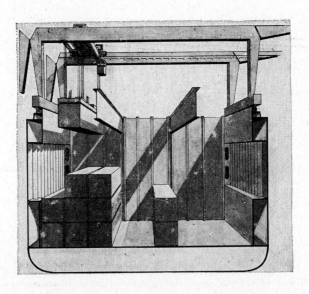

Fig. 15—View looking up from hold in container ship, There are nine container cells in this group, subdivided into three sections of three cells each. Shipboard container crane is lowering container into a cell.

Another problem which arose from the increased use of containers was the design of corner fittings for the purpose of lifting or of locking them together and which would be accepted as standard. The International Standards Organization with headquarters in Geneva, selected the MH-5 corner fitting which is adaptable and will take a standard 30-ton hook, a standard shackle, a twist lock or the special hook designed for container work.

Much research has been carried out on methods of lifting containers which may have off-balance or off-centre loads. The outcome has been the development of electrically controlled automatic self-levelling spreader beams. They are designed to pick up a container and to level it automatically as soon as the weight is taken so that locating containers one on top of another no longer presents any difficulty. These beams are now so efficient that a container can be made to take up an identical angle fore-and-aft to that of the vessel so that the container does not bind in the channelways when being loaded or discharged.

FIG. 16.—Container Ship at Sea.

Naturally the increased use of containers has called for the design of specialised ships to handle them. New types of ships are rapidly being built to carry at relatively high speeds of 23 knots or over not only a large number of containers part of which may be refrigerated, but also other special cargoes. Stowage of large numbers of containers on these ships calls for cellular arrangements such as ten rows athwartships, seven deep below decks with possibly eleven rows or more in the fore and aft direction according to the spaces available. In addition containers may be carried two or more deep above deck as a deck cargo. The containers are held in place by channel-ways and may be locked together by use of conical dogs and the special corner fittings already referred to and as illustrated. Containers are usually built with standard fork-lift pockets for use as needed though it is the practice to land them directly on to the truck or carrier which is to take them on the next stage to their destination.

F

Pallets.—The constant search for improved and more economical cargo-handling techniques is not limited to containers and as an alternative, an earlier method of unitisation has been considerably developed in the past few years. This is the system of handling small units of cargo on pallets with the object of the consolidation of numbers of small packets into larger but fewer unit loads. These pallets are small compared with containers and usually consist of single or double wooden platforms on which the cargo is stowed and which provides pockets or spaces at the bottom for easy handling by fork-lift trucks or machines used both ashore and aboard ship. The size of pallets have been metricated and more or less standardized at (*a*) 1m × 1·20m, and (*b*) 1·20m × 1·60m, though these may vary to suit the particular trade. In the specially-constructed ships the height to which pallets are stowed is 1·8m. These figures closely follow those already given in imperial measure. Pallets may be of metal and used over and over again by the shipper, or if of wood, may be quite easily kept in repair for continuous use or may be considered as expendable as part of the unit load.

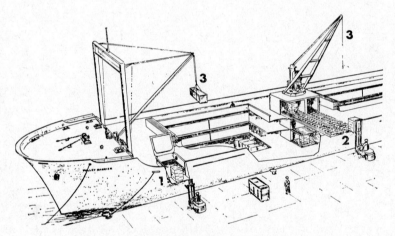

Fig. 17.—Illustrating loading/discharging systems for mechanized handling of pallets:

1. Via Side-ports.
2. Via Pallet Lifts and Roller Conveyors.
3. Via Lifting Gear—Derricks and Cranes.

The full development of the system involves the use of specially built fully mechanised side-port vessels, i.e. vessels with side doors and possibly side hatches through which palletised (unitised) cargo can be passed in either direction and from deck to deck by means of fork-lift trucks. An additional system of roller conveyors and cage lifts or elevators may be installed for moving the pallets not only from deck to deck but to a working platform at the ship's side where the cargo units can be directly picked up by shore based fork trucks. Discharging or loading can also take place by the usual means of the ship's derricks or cranes so that with all of these three methods in operation at one and the same time, an extremely fast and flexible transportation service can be provided. The fork-trucks on board are fitted with side lifters so that the units can be stowed tightly and the cargo is secured by means of wire stay walls and inflatable rubber cushions.

As cargo can be palletised by the shipper at the factory, unitisation by the use of pallets not only reduces the possibility of pilferage but as it involves less handling of shipments during transit there is less likelihood of breakages and a considerable saving in labour costs. The use of pallets makes for speedier turn-round time at the ports and though special ships have been and are being built, the present conventional dry-cargo vessel can be comparatively easily converted to handle this unit pallet system.

Roll-on, Roll-off System.—This system was developed and has proved successful on a number of short Continental runs. In this case, loaded vehicles are driven aboard the vessel through special stern or bow doors and are then properly secured for the passage. On arrival at the port of discharge, the vehicles are released and are driven ashore to their further destination. Train and car ferries operate on the same principle which gives a very rapid turn-round.

Present Trends.—Shore-based handling equipment has endeavoured to keep pace with the container revolution and complex cranes have been developed to speed up the ship to shore movement of even the largest containers. Completely automated computer controlled warehouses have been designed with integral conveyor or lifting equipment for storage and retrieval of a container in a time of less than two minutes.

With the full development of the system, fewer of these fast containerised ships will be needed to carry the cargo available and with the increased speed of loading and discharging, the ships will be in almost continuous service as is the case with the large bulk carriers so that successful operation may finally become a question of the need for adequate maintenance.

At this stage the developments in containers, the ships to carry them and the handling equipment both ashore and afloat, is proceeding rapidly but the roll-on, roll-off system and the use of pallets for certain trades is likely to be extended.

There are other possible developments of which the most interesting is the Lighter Aboard Ship "Lash" concept in which barges of up to 300 tons may be carried aboard a parent ship which is fitted to discharge these loaded barges at the main port. They are then left to be towed elsewhere or otherwise for the cargo contents to be discharged by ordinary means and finally are again loaded with cargo prior to being reloaded aboard the parent ship. This system has the advantage of being simple and as it requires no special port facilities it could be used to serve less developed areas. Interest has also been taken in the building of a vessel with loaded cargo sections which can be floated out, and in which pre-loaded sections can be floated in, which would allow much larger cargo units to be handled.

CHAPTER VI.

BOATS AND INFLATABLE LIFERAFTS

Lifeboats. Types. Parts and Equipment. Inflatable Liferafts. Operation and Equipment. Boat Stations. Launching. Handling in Heavy Weather. Use of Oil. Distress Signals. Boat Covers and Falls. Boats Under Oars. Boats Under Sail.

The Merchant Shipping Life-saving Appliances Regulations 1980. These Rules came into operation as from 25th May 1980, and whilst some of the requirements can only be immediately effective with new construction, any existing alternatives must be approved by the Department of Trade. The following paragraphs only take into account the provisions of these new Rules as they affect the majority of foreign-going cargo and passenger ships. Amongst other things the official recognition of the life-saving properties of inflatable liferafts has been extended and vessels under 500 tons may now dispense with lifeboats altogether if properly supplied with liferafts. Tankers over 3,000 tons having no superstructure amidships are allowed to carry rafts in lieu of the two midship boats. The Rules emphasise a preference for gravity davits for all boats although luffing davits may be used with lifeboats whose all-up launching weight does not exceed $2\frac{1}{4}$ tons.

Number of Lifeboats and Liferafts.—All foreign-going cargo vessels (Class VII) must have enough lifeboats for all on board on both sides of the vessel and in addition liferafts for half that number of persons. There must be at least one 4-knot motor lifeboat with ready-at-hand portable radio equipment while tankers over 1,600 tons must carry two such, one on each side of the vessel. Foreign-going passenger ships must have lifeboats on each side to accommodate half the number of persons she is certified to carry and liferafts for 25 per cent with additional buoyant apparatus for 3 per cent of that number. At least one lifeboat on each side must be a motor lifeboat equipped with both a searchlight and radio. One boat on each side, not more than 28 feet in length and which may be a motor lifeboat must be rigged for immediate use as an emergency boat.

Lifeboats.—Lifeboats are built to D.o.T. specifications and for foreign-going vessels must not be less than 24 feet in length. They must have rigid sides and be strong enough to withstand lowering against a 15-degree list and a trim of 10 degrees when they are 25 per cent overloaded. Afloat they must have ample stability and freeboard. If a fully-laden boat is flooded, it must float with the tops of its gunwales clear of the water. Internal buoyancy in the form of watertight air-cases made of yellow metal not less than 18 ozs. to the superficial foot or equally buoyant material is fitted to occupy at least one-tenth of the internal capacity of the boat. No loaded lifeboat shall ever exceed 20 tons in weight and allowing an average of 165 lbs. per person, the maximum number of persons must never exceed 150.

If a boat is deemed to carry more than 60 persons it must be either a motor-boat or be fitted with other approved mechanical means of hand propulsion, but if for more than 100 persons it must be fitted with a motor and comply with the same requirements as a motor lifeboat.

All thwarts and side seats must be fitted as low in the boat as is practicable and bottom boards have to be fitted so that the thwarts will be not more than 2 feet 9 inches above them.

Lifeboat Markings.—The name of the ship and the Port of Registry is required to be shown in each lifeboat. In addition, the dimensions of the boat, its cubic capacity, and the number of persons it is designed to carry must also be cut into the inboard side of the stem or gunwale. The figures must not be painted over, but should be kept scraped and varnished in order that they will be easily read by the Surveyor.

To Find the Cubic Capacity.—Take the outside length in feet and multiply it by the outside breadth and the inside depth, then multiply the result by 0·6. This will give the capacity of the boat in cubic feet or it can otherwise be found by calculation based on Simpson's Rules.

Note.—The shape of a boat in relation to a cubic block of the same dimensions is known as the coefficient of fineness and is expressed by a decimal which in this case is 0·6.

In the case of a metal or fibreglass boat the coefficient is increased to at least 0·64 and in any case the resulting capacity must be reduced by the volume occupied by the machinery in the case of the mechanically-propelled and motor-boats.

Number of Persons Boat will Carry.—As each person in a boat is allowed 10 cubic feet of space, it is only necessary to divide the cubic capacity of the boat by 10. This will give the number of people she will carry providing there is seating room and that there will be no interference with the working of the boat when all hands are wearing lifejackets.

Numbering Boats.—Numbers always start on the starboard side forward. Opposite No. 1 on the starboard side we have No. 2 on the port side, then No. 3 would be found abaft No. 1 on the starboard side and No. 4 abaft No. 2, and so on. When there are two boats to one set of davits the lower one is also marked alphabetically. For instance, the top one might be No. 6, then the bottom one would be No. 6A.

Certificated Lifeboatmen. REGULATION 47.—**Lifesaving Appliances Regulations** 1980.—Each lifeboat or liferaft which forms part of the Statutory Life-saving Appliances of a sea-going passenger ship shall have a number of certificated lifeboatmen, and these shall be not less than the number specified in the following table.

Prescribed Complement of Lifeboat or Liferaft.	*Minimum Number of Certificated Lifeboatmen.*
Less than 41 persons	2
From 41–61 persons	3
From 62–85 persons	4
Above 85 persons	5

Lifeboatman's Certificates are issued by, or under the authority of the D.o.T. and any member of the crew (including mates) may act as a lifeboatman, provided he holds such a certificate. *See* page 11.

Emergency Signal.—The signal for summoning passengers to the assembly stations shall be a succession of more than six short blasts followed by one long blast on the whistle or siren.

Musters.—In passenger ships, musters of the crew shall take place weekly and in cargo vessels fortnightly, to ensure that all members of the crew understand and are drilled in their respective duties in the event of an emergency.

TYPES.

Various types of boats will be found in different classes of vessel, but the D.o.T. requirements are the same in all cases. It is mostly in the method of construction where they differ.

Clincher Built.—A boat with the edges of the planks overlapping is said to be "clincher built".

Carvel Built.—When the edges of the planks are butted together to form a smooth surface the boat is said to be "carvel built".

Diagonal Built.—These boats are "carvel built" but the planking (in two thicknesses) runs obliquely across the timbers. This construction is recommended for all wooden boats carrying more than 60 persons and compulsory for larger wooden boats carrying more than 85 persons.

Hand-Propelled Boat.—To do away with the use of oars these boats are fitted with a propeller worked by simple gear-wheels. The gearing is operated by a number of vertical levers placed between the thwarts in two rows. By means of a small reverse lever the coxswain is able to go ahead or astern, and the speed of the boat is about $3\frac{1}{2}$ knots in smooth water for $\frac{1}{4}$ mile.

The equipment of these boats is the same as for other lifeboats except that a 50 per cent reduction in oars is permitted, and additional buoyancy tanks are required to make up for the weight of the gear. No mast or sails are required.

Jolly-boat.—The average tramp steamer always carried a jolly-boat which was invariably clincher built. It was used for such purposes as rowing the master ashore, running lines to buoys, painting the ship's side, etc. They had no internal buoyancy tanks and were not counted as lifeboats.

Motor-boats.—Motor-boats which form part of the statutory life-saving appliances must comply with the same requirements as a lifeboat except that they need not carry a mast or sails or more than half the complement of oars. Two boathooks must be provided and additional buoyancy tanks are also required.

They must have a reliable and easily-started compression-ignition (Diesel-type) engine, able to work astern and covered by a fire-resistant casing securely fixed above a metal tray designed to preclude the spread of oil.

Fuel for 24 hours' running must be at all times kept in the boat and the speed when fully laden in smooth water shall be not less than four knots for cargo vessels but not less than 6 knots for passenger vessels.

They must carry two 1-gallon fire extinguishers capable of fighting oil fires and a full bin of sand with a scoop for spreading the sand.

The two motor boats required to be carried by a passenger ship must be fitted with radio telegraph equipment in a cabin large enough to accommodate both the equipment and the person using it. Each boat must also be fitted with a searchlight and the source of power be sufficient to maintain efficient service of either or both installations for a period of 6 hours.

Steel Boats.—This type has become popular on account of their fire-resisting qualities. They are constructed with two single sheets of steel stamped into shape and riveted together. The join is in the centre line of the boat. The buoyancy of a steel boat must be not less than that of a wooden one of the same cubic capacity, so the volume of the watertight air-cases has to be increased at the rate of $1\frac{3}{4}$ cubic feet for each cwt. of steel.

Glassfibre Boats.—Glassfibre is light, strong, free from corrosion or rotting and lends itself to moulding processes by which buoyancy can be built into the hull. It is unaffected by extremes of temperature and most damage repairs can be quickly and effectively carried out so that maintenance costs are almost negligible.

Aluminium Boats.—Boats constructed of aluminium alloy must be equally as strong as wooden boats and though it has the advantage of being a light material not affected by exposure to the elements, care must be taken to ensure that no galvanic action takes place between dissimilar metals being used for fittings. Paint on an aluminium surface must have a zinc base and on no account must lead-based paint be used.

Inflatable Boats (Boti Boats).—These are carried in some ships such as supply boats and tugs. They must be at least 3·8 metres long and must carry at least six persons.

PARTS OF A WOODEN BOAT.

Apron.—Part of the stem and stern posts. The material used is British oak, teak or ash.

Bilge.—The lower part of the rounded side which the boat would rest on if allowed to fall over on her side.

Block Span.—A light, wire span stretching between the two lower purchase blocks to prevent the falls twisting while they are "rounded up".

Bottom Boards.—A framework of light boards is fitted to the boat's bottom to protect the boat and keep the feet dry when water is lying there.

Bow Sheets.—The raised platform in the bows.

Buoyancy Tanks.—Only tough non-corrosive metals are used for building these tanks. Brass, copper, muntz or yellow metal are the most popular. In weight it must be at least 18 ozs. per square foot of surface.

Chain Plates.—Eyebolts fixed to the gunwale, and to which the shrouds or backstays are made fast.

Cleading.—The wooden casings at the side of the boat behind which the internal buoyancy tanks are fitted.

Cleats.—Four cleats are fitted inside the gunwale for "turning up" the sheets when sailing.

Deadwood.—A piece of wood used in the structure of the forefoot and made from British oak or ash.

Forefoot.—The rounded part where the stem joins the keel.

Garboard Strake.—The planks on each side next to the keel.

Gudgeons.—A sort of socket on the stern-post into which the pintle on the rudder fits.

Gunwale.—The top edge of the boat into which the crutches are shipped. They are usually about 5 inches deep and made of either American elm. teak, oak or ash.

Keel.—A keel must be all one piece and either of American oak or elm or Australian spotted gum.

Keelson.—A strengthening piece fitted to the top of the keel. The mast step is secured to it.

Life Rings.—Rings fitted at intervals of 2 feet around the sides of the boat, on the under side of the rubber. To these rings the grab-lines are made fast.

Lifting Hooks.—Those to which the blocks are hooked. The points must face amidships, and they are secured down through the keel and keelson by means of keel plates and bolts.

Lower Thwarts.—Seats which fit in between the side casings for passengers to sit on, and for oarsmen to brace their feet on. They are removable, so that injured people may be laid in the boat.

Mast Clamp.—A hinged semi-circular iron band with a retaining pin which keeps the mast in an upright position.

Mast Step.—A wooden socket on the keelson into which the heel of the mast is stepped. The step may be of metal.

Pintles.—The prong arrangement on the rudder which fits into a gudgeon or socket on the stern-post, thereby forming what is practically a hinge.

Planking.—As a rule larch or yellow pine is used for planking.

Plug Holes (or Draining Holes).—At least one plug hole will be found towards the after end of the boat. When two are fitted they may be placed on each side of the keel, or one of them forward.

Roove or Rosebur.—Small washer fitting over a copper nail for riveting boat planks together.

Rubber.—The half-round wooden moulding at the bottom of the gunwale which is meant to act as a fender.

Sheer Strake.—The planks on each side immediately under the gunwale.

Stem Band.—A metal band running from the stem head to a little abaft the forefoot, which it both protects and strengthens.

Stem-Post.—Is made from British oak, teak or Australian spotted gum.

Stern-Post.—Is made from British oak, teak or Australian spotted gum.

Stern Sheets.—The raised platform in the stern.

Thwarts.—The seats which extend from one side of the boat to the other.

Timbers.—These may be likened to ribs. They are spaced not less than 6 inches apart and extend from side to side in one piece. The usual material is rock elm, oak, ash or teak.

GENERAL LIFEBOAT EQUIPMENT.

Baler.—A galvanised iron baler not less than 8 inches in diameter must be attached to the boat by a lanyard and be kept ready for use.

Bilge (or Keel) Rails.—Fitted outside to the bilge together with lines secured round the boat to enable persons to cling to the boat should it be upturned.

Boarding Ladder.—To give access to the boat from the water.

Boathooks.—One boathook must be provided but two are required in the case of motor-boats.

Buckets.—Two galvanised iron buckets to hold about two gallons must be attached to the boat by lanyards.

Charts.—An approved set of charts must be carried in a waterproof wallet, together with Azimuth tables, protractor, writing-paper, pencil and eraser.

Compass.—An approved "liquid" type of compass must be fitted in a proper binnacle with means of illumination. The card must be not less than 4 inches in diameter and may be luminised or provided with an oil lamp capable of burning for 10 hours.

Cover.—A cover of a highly visible colour capable of protecting the occupants against injury by exposure.

Crutches.—One set and a half of crutches attached to the boat by lanyards. A grommet or a crutch must be provided for the steering oar.

Dipper and Drinking Vessels.—At least one dipper with lanyard must be provided and three rust-proof drinking-vessels of which one must be graduated in $\frac{1}{2}$, 1 and 2 ozs.

First Aid Outfts.—An approved outfit must be carried and includes a drug in tablet form (omnopon) for the purpose of deadening pain.

Fishing Line.—Together with 6 hooks.

Grab Line.—The line which must be becketted round the outside of the boat and of which the loops should be just clear of the water when the boat is loaded.

Hatchets.—Two hatchets must be provided, one at each end of the boat and must be attached by suitable lanyards.

Heaving Lines.—Two light buoyant heaving lines must be carried.

Inventory.—An inventory must be kept of all the equipment carried in the boat.

Jack-knife.—This must be fitted with a tin opener and be attached to the boat by a lanyard.

Keel Rails.—See Bilge Rails.

Lamp and Matches.—A lantern of an approved type must be kept trimmed with sufficient oil to burn 12 hours and two boxes of matches (in watertight box).

Locker.—All boats must be provided with a suitable locker for the stowage of small equipment.

Manual Pump.—This must be fitted for bilge pumping purposes.

Mast.—The mast must be not more than two-thirds the length of the boat, fitted with mastband and wire stays to be secured to chain plates on the gunwale.

Oars.—A full single-banked complement and two spares; also a steering oar which should be one foot longer than the others and with the blade painted white. The blades of the oars are always laid forward, with the exception of the steering oar which lays aft.

Oil Bag.—Small canvas bag (or other approved appliance) for distributing oil on the sea. It is partly stuffed with oakum and arranged for attachment to the sea anchor.

Oil Can.—Contains one gallon of vegetable, fish or animal oil for use with the oil bag.

Painters.—There must be two painters of sufficient length (at least 20 fathoms); one must be secured to the forward end of the lifeboat with strop and toggle so that it can be released from the boat and the other must be secured to the stem and ready for use.

Plugs.—Two plugs for each hole are required and must be attached with lanyards or chains. Plugs may be dispensed with if proper automatic valves are fitted.

Pyrotechnic Signals.—Four hand-held parachute rocket distress signals and six hand-held flare distress signals must be carried in watertight cases. The signals must be clearly labelled with instructions as to their use.

Rations.—The following must be provided for each person the boat is deemed fit to carry:—

(a) At least 6 pints of fresh water or, alternatively, 4 pints with de-salting apparatus able to provide at least another 2 pints.
(b) At least 450 grams of biscuits.
(c) At least 450 grams of barley sugar.
(d) At least 450 grams of sweetened condensed milk of first quality.

Rescue Signal Tables.—It is compulsory to provide a copy of the Ministry of Transport Rescue Signal Tables as published by H.M.S.O.

Rudder and Tiller.—The rudder must be attached to the boat by lanyards to prevent it being lost if accidentally unshipped. It is hung in position by means of pintles fitting into gudgeons on the stern-post.

Sails.—At least one good sail (orange-coloured), and if the boat is over 25 feet in length a jib must also be fitted. For identification purposes the sail must carry in black lettering the first and last letters of the vessel's name and the number of the boat.

Sea Anchor.—Must be of best flax canvas; not less than 27 inches in diameter at its mouth and 4 feet in length with the diameter tapering to 5 inches at the other extremity. The bridle is at least 2 feet 6 inches in length and a hawser at least three to four times the length of the boat must be provided together with a tripping line 2 fathoms longer than the hawser.

Signalling Mirror and Signalling Whistle.—The mirror must be of an approved type for daylight signalling.

Smoke Signals.—Two buoyant smoke signals are for day-time use and must be capable of giving off a volume of orange-coloured smoke.

Stowage.—All items not kept in the locker shall be lightly lashed within the boat. The boathook, however, must be kept free for fending-off purposes.

All boats must be fully equipped and provided with rations before the ship goes to sea and these must remain in the boat while the ship is at sea.

Torch.—An electric torch must be carried of a type suitable for morse signalling and supplied with two spare batteries and two spare bulbs.

Traveller.—This is fitted on the mast to take the yard and is provided with halliards for hoisting the sail.

Water Containers.—These must be of an approved size and design and the water must be frequently changed to ensure that it is always clean and fit for drinking. Additional fresh water should be provided if at all possible.

Water-tight Tanks.—All food must be carried in water-tight containers properly labelled and stowed in water-tight tanks. Two tank keys are usually provided and consist of flat pieces of metal which fit into the slot on the screw-caps of the tanks.

Yard.—A suitable yard to which the head of the sail is bent.

INFLATABLE LIFERAFTS.

The raft may accommodate from 6 to 25 persons and must remain afloat if only half the air compartments are inflated. It must be capable of being righted by one man should it inflate in the inverted position. The floor must be waterproof and capable of inflation by hand to insulate against the cold. It must withstand any temperature from $-30°$ C. to $+66°$ C. and must survive exposure to any weather, afloat at sea for 30 days.

Description.—Such liferafts or dinghies are either round or oval in shape, made of rubberised nylon material and become buoyant on inflation by the escape of carbon dioxide gas from a small cylinder attached to the floor. There are two main buoyancy chambers each capable of supporting the occupants, and in addition there are usually two inflatable arches supporting a canopy or tent with openings at each end, provided with a strap or ladder for climbing on board from the water. The opening can be closed and the floor can be separately hand-inflated as needed to ensure protection and warmth. The raft must be clearly marked with the maximum number of persons it is designed to carry and is fitted with a painter and a line becketed round the outside. There is also a lamp worked by a sea-activated cell, fitted to the top of the cover for the purpose of locating the position.

Initially the liferaft is fitted into a valise for easy handling and to withstand hard wear, but the total weight must not exceed 400 lbs. unless it can be launched without lifting. If inflation takes place in an inverted position, it can be readily righted and water-pockets then ensure conditions of stability.

Operation.—
(1) See that the operating cord is secured to a strong point on the vessel.
(2) Throw the raft in its valise overboard.
(3) Pull on the operating cord until it is tight and then give a sharp tug to start inflation, which should be completed within 30 seconds.

Equipment.—
(1) Two sea anchors; one permanently attached and one spare with line.
(2) One baler, sponge, jack-knife for every 12 persons and one topping-up pump or bellows.
(3) One puncture repair kit for emergency use.
(4) Two paddles and one rescue quoit attached to at least 100 feet of line.
(5) Instructions on how to survive in liferaft.

The following completes the full equipment and may be supplied as a special emergency pack:—

(6) Two parachute distress signals and six hand flares.
(7) One electric signalling torch (with battery and bulb), one daylight-signalling mirror and one signalling whistle.
(8) Fishing line and six hooks.
(9) $1\frac{1}{2}$ litres fresh water per person and usual graduated drinking vessel.
(10) 340 grams non-thirst food and 170 grams barley sugar for each person.
(11) Three safety tin-openers; six seasickness tablets per person and first aid outfit.

Note.—(*a*) These items are not required in the case of vessels confined to smooth or near coastal waters.

(*b*) A Log Card is provided with each liferaft or dinghy and certified entries are made when supplied and at each servicing, which must be at intervals of not more than 12 months.

Inflatable liferafts (first made compulsory for certain fishing vessels as from 1st October, 1956) are now compulsory for all fishing vessels over 50 feet; there must be sufficient capacity to accommodate all persons on board and if the number is 13 or over, there must be at least two such liferafts.

Other types of vessels moving in smooth or partially smooth waters and vessels restricted to short voyages in near coastal waters carry these liferafts in lieu of buoyant apparatus. Home trade passenger vessels and coasters must carry inflatable liferafts as part of the life-saving appliances and foreign-going cargo vessels (other than tankers) should carry one or two to accommodate half the ship's complement.

BOAT PROCEDURE.

Davits.—The old type radial davits which were cheap and simple to install are now almost obsolete and are not permitted by the D.O.T. in any new construction. Launching a boat with this type of davit which necessitated the boat being swung out one end at a time, was a difficult and dangerous operation, particularly in heavy weather.

The present Rules demand a high standard of equipment and emphasise a preference for gravity davits for all boats although luffing davits may be used with lifeboats whose all-up launching weight does not exceed $2\frac{1}{4}$ tons. In general a two-man crew should be able to launch a boat under gravity alone when it is fully equipped and holding a full complement of persons against a list of 15 degrees and a 10-degree trim. In any case where it might be unsafe to do this, the davits must be marked with a 6-inch band of red on a white ground.

All the systems are given in the following pages.

Boat Winches.—These winches should be sited so that the operator can watch the boat at all times whilst being lowered or hoisted and this must be so in the case of emergency boats. The two winches must be arranged to lower at the same rate of between 60 and 120 feet per minute, the automatic brake being set to control the speed. The winches should be able to hoist the boat with its gear and a two-man crew at least 60 feet per minute.

Clearing Away Boat.—Strip cover, strongback and spreaders. Take cover off the boat falls and see they are clear for lowering, especially if tied with a light seizing well clear of the deck.

Lifelines are usually fastened in the same manner to either the davit head or the top block: make sure they are hanging clear. These, by the way, must be long enough to reach the water when the ship is at her lightest draught and listed over, away from the boat, at an angle of 15 degrees.

See also that any lashing on the oars are cast adrift.

When the painter has been passed forward outside the davit, and the plug shipped, the boat is all ready for swinging out.

Finding the Plug.—It is not always easy to find the plug-hole, especially at night, so for this reason an arrow is sometimes cut into the gunwale, abreast of the hole, to indicate where it will be found.

Stretching the Painter.—Never stretch a painter too tightly: always allow plenty of slack. If this is not done the boat may be "hung up" when half-way down the ship's side, and when in the water she will draw away from the ladder, owing to the weight on the painter.

GRAVITY DAVIT.

Gravity Davits (see Figure 4).—Although the method of operating the various types of gravity davits may differ slightly in some cases, the general principle is always much the same as the one given here.

Lowering.—Release gripes by striking slip link with bar provided for the purpose. Lift control handle gently and allow both cradles and boat to run down trackway until cradles come to rest on the buffers. Carry on lowering until pendants draw boat close to ship's side, abreast of the embarkation deck then make fast tricing gear. Note: Tricing gear prevents the boat swinging violently away from the ship's side when the pendants are slipped.

When the complement of passengers are on board, slip both pendants and ease boat off from side with tricing gear. Note: If the alternative method for tricing gear is used it is left in place made fast, the falls rendering through it. When near the water's edge, lower boat at highest speed to ensure that there will be plenty of slack fall when unhooking blocks.

Hoisting.—Switch on power by means of power switch and regulate speed of hoisting by working the speed control wheel. When high enough, make pendants fast and pass gripes over boat. Keep hoisting at speed. Boat will automatically hook on cradles and both boat and cradles will hoist up trackway until the limit switch cuts out the power. Note: Keep an eye on the pendants in case they become wedged between the boat and the cradle.

Boat is now in position for connecting gripes and screwing them tight. Should it ever be necessary, the boat can be hoisted by means of the handle provided for the purpose.

Side Ladders.—All vessels must carry an approved side ladder at each set of davits, and these must be long enough to reach the waterline when the ship is at her lightest sea-going draught and listed to 15 degrees either way. One ladder for each two sets of davits in cargo ships.

Lights when Lowering.—In passenger ships provision must be made to illuminate the davits, the interior of the boat and the boats while lowering and immediately after being launched. There must be a separate source of power above the bulkhead deck capable of

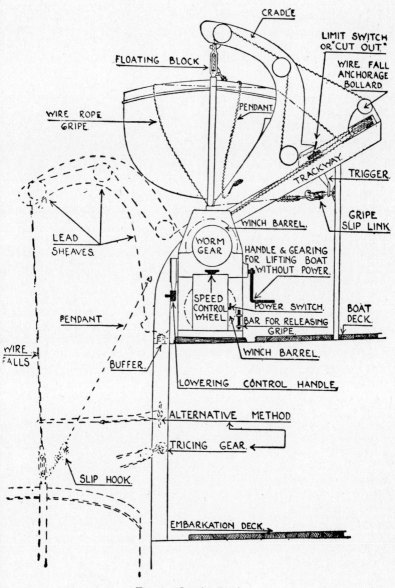

FIG. 4.—Gravity Davit.

supplying this illumination.   In all other vessels lighting must now be supplied at the davit positions.

Lying Alongside.—To lie alongside for even a few minutes in bad weather is sometimes a dangerous business, but is often necessary if waiting for more passengers.

One of the principal dangers, if not the worst, is that the painter being fast to the stemhead, it will bind the boat's bow into the ship's side which is not always desirable.   Freedom to manoeuvre the boat close into the ladder, or out again quickly, is a big advantage, and can only be done properly if the painter is made fast to the first thwart.   Another danger to guard against is the disengaged lifeboat falls swinging about.   They may either damage occupants of the boat or the boat itself.

Getting Away from a Weather Side.—When bound by wind and sea to a weather side the painter should be passed aft, and a pull can be taken on it from there while the bow is shoved off.

Holed Below Waterline.—It may happen that a boat will sustain some damage below the waterline when in the vicinity of a wrecked or foundering vessel.   In cases of this kind a large inflow of water can often be stopped by spreading the jib, or even a blanket, under the hole.   Secure it in place with the halliards, led underneath the boat if necessary.

Awaiting Help.—Through the agency of wireless, help is soon forthcoming to a crew after a vessel has been abandoned, and for this reason boats should keep together as much as possible until further orders.   This is best accomplished by riding to a sea anchor and the use of oars when necessary.

Putting Out a Sea Anchor.—See the tow line and tripping line clear for running, then pay out over the bow and let the boat drift from it, at the same time keeping the boat's head up to wind and sea with the oars.   Reeve end of tow line through ring on stem, and make fast around the forward thwart or the mast.   If likely to chafe anywhere, the line must be parcelled over with any material which happens to be handy.

Improvised Sea Anchor.—If the sea anchor is not sufficient to keep the boat's head up to the sea, lash the mast, sail, and some oars together, and allow them to drift ahead with the painter bent to the middle.

Putting Out Oil Bag.—To prevent water breaking over the boat when riding to a sea anchor, bend the jib halliards to the oil bag, which should be about half or two-thirds full, and pay out the full length of halliards.   In cold weather the oil must have time to diffuse, so it is better to make the oil bag fast to the sea anchor.

Attracting Attention.—To attract the attention of passing vessels, hoist a blanket or any article which is likely to be seen at a distance.

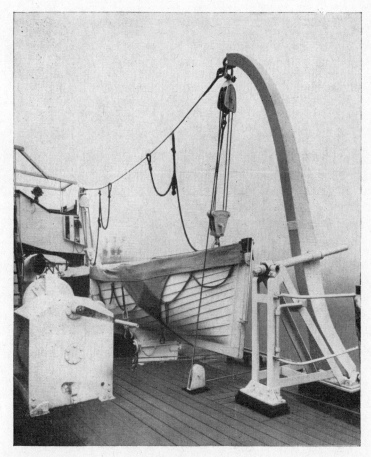

Fig. 5—Welin 'Lum' Davit

Note:—This is a hand cranked luffing davit suitable for wire rope falls and winch operation (as shown) or hand tackle. Luffing davits of this style are positioned at the end of the boat which is chocked on deck. The action of the boat tackle provides a 'compensation' effect which lifts the boat keel over the ship's curtainplate to outboard.

At night, the S.O.S. may be flashed by dipping the lamp in and out of a bucket, and a red light or any kind of flare may be burnt also.

Lighting Pyrotechnic Distress Signals.—By means of a tape at the top of the signal, expose the igniting composition; now remove the striker from the bottom of the signal by pulling downwards on the tape at that end; finally, strike the exposed igniting surface firmly with the striker. Hold the signal upright to leeward and away from the body. The bright red flare is for the purpose of attracting attention.

The above is the older method and is being superseded by semi-automatic devices which operate either by a twisting action or by releasing a loaded spring. In any case the procedure is always clearly indicated.

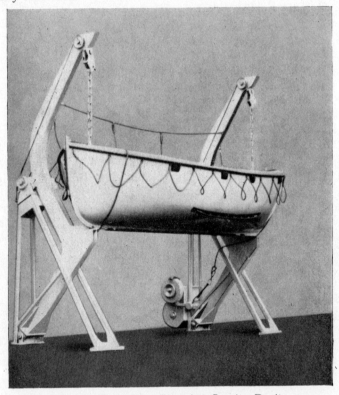

FIG. 6—Welin Pure Pivotting Gravity Davit

NOTE:—This is a modern davit which is entirely self-contained, and no rope leads appear on the ship's deck. The boat winch is mounted on the davit frame as shown, and the rope falls are reeved to allow an 'endless' effect, i.e. there is no 'standing part' to be clamped to the davit frame. The keel height of the stowed boat allows complete access to the ship's side, this height being in the region of seven feet.

Shelter in Boat.—When a vessel is abandoned in winter time or during bad weather, the strong-back and cover should be kept in the boat to provide shelter for the crew and passengers. The type of cover laced to studs under the rubbing-strake is best adapted to this purpose. Even in a crowded boat, if there is no room for the strong-back, the cover at least should be taken.

By placing male passengers on the upper thwarts and side casings, more shelter will be given to women and children on the lower thwarts and bottom boards.

Extra Provisions.—As many extra provisions as possible should be taken in a boat—particularly water. Both sugar and jam are valuable for creating bodily heat, and in this connection it is as well to remember blankets—if time permits.

To Alleviate Thirst.—To promote the flow of saliva keep a button in the mouth. The drinking of salt water is dangerous, and usually fatal, but keeping the clothing wet gives great relief.

Spare Gear.—A certain amount of spare gear is required to be carried as equipment, but it is as well to be prepared for emergencies and include serviceable odds and ends which do not take up much space.

STEERING BY STARS

All seamen should be able to steer an approximate course by means of stars so that, in the event of a lifeboat compass becoming damaged, they will not be altogether helpless as a result. Fig. 7.

The important groups for this purpose are the Big and Little Bear which contain the North Pole Star, and the Southern Cross which indicates where to look for south.

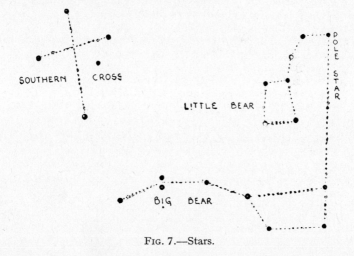

FIG. 7.—Stars.

North Star.—A line drawn through the two end stars of the Big Bear points to the tail star of the Little Bear which is the North Pole Star itself. The Big Bear can be seen as far as 30° S., so an approximate bearing of the North Star can be obtained in those atitudes.

Southern Cross.—A line drawn lengthwise through the top and bottom stars of the "Cross" for a distance of four lengths, indicates where to look for South.

The Southern Cross cannot be seen much farther than 23° N.

Method by Watch.—Hold an ordinary watch in a horizontal position, with the face uppermost and the hour-hand pointing to the sun.

To get the exact direction of the sun, when it is high, it is better to place a match vertically against the rim of the watch, opposite the point of the hour-hand, so that a shadow falls on the hand itself.

Midway between the hour-hand and the figure 12 indicates south.

In the southern hemisphere, south of the sun, north will be the compass point indicated.

HANDLING A BOAT IN HEAVY WEATHER.

The curling crest of a breaking sea should at all times be avoided either when rowing against, or running before a heavy sea. Failure to do so may result in a boat up-ending or broaching to. By regulating the speed, a sea may be allowed to break either ahead or astern of the boat.

Rowing Against a Sea.—Speed is an important factor in the management of a small boat in breaking seas. For instance, an increase of speed will carry a boat through or over a breaking crest, whereas, without forward movement, there is a likelihood of the boat up-ending and capsizing. But care must be taken not to have too much speed on the boat, or she will fall suddenly and heavily when the crest has passed. By retarding speed in sufficient time a sea may be allowed to break ahead of the boat.

Running Before a Sea.—This is the most dangerous position for a boat and calls for a good deal of judgment and skill. The danger arises because the motion of the boat is in the same direction as the sea, and while the stern is lifted in the air, the bow is depressed and meets with the resistance of solid and practically stationary water. With skilful steering the boat may run on the front of a sea for some distance, at a high speed, until finally the wave breaks and is spent, or she may bury her bow too far and the crest, still pressing her onward, will force her stern either to one side or the other, causing her to broach to and capsize, or throw her end over end. To obviate this danger the oars should be backed to allow the sea to pass, then immediately start rowing ahead again, travelling on the back of the wave to take advantage of the speed to be gained.

Trim.—From the foregoing it will be seen that the best trim is deepest by the stern, to prevent that part being too easily thrown to one side, or lifted too high.

Steering.—When running before a sea, a steering oar over the stern should always be used in place of the rudder.

Landing Through a Surf.—Two methods, each dependent on the size and shape of a boat, may be employed when landing through a surf. For instance, with ordinary lifeboats, a drogue in the form of a sea-anchor may be towed astern. This will check the speed of the boat when a breaking sea has to be avoided. Good progress can be made by alternately working the drogue and the oars.

If a boat is small, or if it has a square stern, the bow should be turned to seaward, so that she can be taken in stern first. To help the process of backing oars, half the oarsmen should sit facing forward so that a more hearty pull can be given to propel the boat towards the beach.

Beaches.—As a general rule high land is a fairly sure indication of deep water near the shore, and this would probably give us a *steep* beach, but low lying land is invariably a sign of shallows inshore, and this would indicate a *flat* beach.

Flat Beaches.—Very often a flat beach or shallows of this type will extend a good way out to sea—a matter of miles—and it is on the fringe of those shallows where the breaking seas will be heaviest and most dangerous, therefore sail should be lowered before reaching it. As the water shoals, so does the danger lessen. Close to the shore the force of the sea is spent, and it will be safe to run the boat on the beach end on. Each wave will carry her farther up, and in this the boat may be assisted if the crew jump out to lighten her, and prevent her being carried off again by the backwash.

Steep Beaches.—As the heaviest fall of water is on the shore itself, a boat on the back of a sea may row or sail right in without losing speed, until she touches the shore, then her stern should be swung round until she is broadside to the surf. The strong force of broken water striking such a large area will quickly sweep the boat bodily up the beach whereas, when end on, the area exposed to the sea is so small that water passing the boat, and breaking on the shore would rush back down the beach and quickly carry her off again, placing her in a very dangerous position.

Unknown Beaches.—To attempt a landing on an unknown beach at night is extremely dangerous. It would be better to lay to all night and at least wait for daylight in the morning. Either that or fire off a distress signal, which might bring help in the shape of a lifeboat or a tug.

THE USE OF OIL.

The Board of Trade desire to call attention to the following information, which has been published by the Admiralty in their

Sailing Directions, on the Use of Oil for Modifying the Effect of Breaking Waves.

Many experiences of late years have shown that the utility of oil for this purpose is undoubted, and the application simple.

The following may serve for the guidance of seamen, whose attention is called to the fact that a very small quanity of oil, skilfully applied, may prevent much damage, both to ships (especially the smaller classes) and to boats, by modifying the effect of breaking seas.

The principal facts as to the use of oil are as follows:—

(1) On free waves, i.e., waves in deep water, the effect is greatest.

(2) In a surf, or waves breaking on a bar, where a mass of liquid is in actual motion in shallow water, the effect of the oil is uncertain, as nothing can prevent the larger waves from breaking under such circumstances; but even here it is of some service.

(3) The heaviest and thickest oils are most effectual. Refined kerosene is of little use; crude petroleum is serviceable when nothing else is obtainable, but all animal and vegetable oils, such as waste oil from the engines, have great effect.

(4) A small quantity of oil suffices if applied in such a manner as to spread to windward.

(5) It is useful in a ship or boat, both when running or lying to, or in wearing.

(6) No experiences are related to its use when hoisting a boat up in a seaway at sea, but it is highly probable that much time and injury to the boat would be saved by its application on such occasions.

(7) In cold water, the oil being thickened by the lower temperature and not being able to spread freely, will have its effect much reduced. This will vary with the description of oil used.

(8) The best method of application in a ship at sea appears to be—hanging over the side, in such a manner as to be in the water, small canvas bags, capable of holding from 1 to 2 gallons of oil, such bags being pricked with a sail needle to facilitate leakage of the oil.

    The position of these bags should vary with the circumstances. Running before the wind they should be hung on either bow, e.g. from the cathead, and allowed to tow in the water. With the wind on the quarter the effect seems to be less than in any other position, as the oil goes astern while the waves come up on the quarter.

    Lying to the weather bow, and another position further aft seems the best places from which to hang the bags, with a sufficient length of line to permit them to draw to windward, while the ship drifts.

(9) Crossing a bar with a flood-tide, oil poured overboard and allowed to float in ahead of the boat which would follow with a bag towing astern, would appear to be the best plan. As before remarked, under these circumstances the effect cannot be so much trusted.

On a bar with the ebb-tide it would seem to be useless to try oil for the purpose of entering.

(10) For boarding a wreck it is recommended to pour oil overboard to windward of her before going alongside.

The effect in this case must greatly depend upon the set of the current and the circumstances of the depth of water.

(11) For a boat riding in bad weather from a sea anchor, it is recommended to fasten the bag to an endless line rove through a block on the sea anchor, by which means the oil is diffused well ahead of the boat, and the bag can be readily hauled on board for refilling if necessary.

(12) Towing a vessel in a heavy sea, oil is of the greatest service, and may prevent parting the hawser, if distributed from the towing vessel forward and on both sides; if used only aft the tow alone gets the benefit.

ANSWERS TO DISTRESS SIGNALS.

The following signals are used from the shore in connection with the life-saving services in Great Britain and Northern Ireland:

Signal	Meaning
Orange smoke signal. or white star rocket. Three single signals fired at about 1-minute intervals. or exploding white star rocket. Three single signals fired at about 1-minute intervals.	You are seen. Help is coming.

Landing Signals for Guidance of Small Boats or Persons in Distress.—Signals to be made in order to assist crew of a wrecked vessel to land along the coast of the United Kingdom:

Signal.	Meaning
Up and down movement of a white flag, white light or flare or the arms or a green star rocket. Morse code letter K (— · —), given by light or sound.	Land here.
NOTE.—If a second steady flag, light or flare is shown at a lower level	Line the two up and come in on this line of approach.
Side-to-side movement of a white flag, light or flare or of the arms fully extended or a red star rocket. Morse code letter S (. . .), given by light or sound.	Landing dangerous.

1. Side-to-side movement of a white flag,
   light or flare.
2. Placing the flag or light in the ground.
3. Carrying a second flag or light in the
   direction to be indicated.

or

1. Red star rocket fired vertically.
2. White star rocket fired pointing to a better
   landing place.

Landing dangerous. Go
in direction indicated
until the " land here "
signal is given.

BOATWORK IN GENERAL.

Towing.—Always use a short towline when being towed, and always tow a spar with the heavy end foremost.

Towing Alongside.—Never attempt to tow alongside a vessel with a boat rope or painter made fast near the stem head. It can't be done.

Always take a turn around the first thwart and hold the end (or bight) in your hand, so that it may be slipped at a moment's notice if necessary. A boatrope in this position allows the bow to swing free in any direction, whereas, if made fast forward, the bow would be bound in tight to the ship's side and steering would be impossible. A lazy painter is always secured in the same position.

Running Lines.—When running a line to a buoy or quay, the eye of the mooring line should be laid in the bows, and there should be enough slack to reach the bollard or ring where it is to be made fast. A small coil should also be taken in the boat, so that it can be paid out if required.

When running a line to windward or against the tide, it is as well to take the whole coil in the boat, make fast at the appointed place, and drop down with the tide to the ship.

"Smart Landings".—To bring a boat alongside a vessel or landing steps with a bump, and end on, is very bad form. It is just as easy to take the way off the boat and fetch up parallel with the landing steps or pilot ladder.

Accommodation Ladders.—Good judgment is required to take a boat alongside an accommodation ladder when a swell is running. The crew must be constantly on the alert to prevent the boat getting under the platform.

Alongside with Mast Up.—Lying alongside a vessel or quay with a mast up in bad weather is simply asking for trouble. Always unship it before going alongside.

Mooring Boats.—When a lifeboat is in use and it is to be left in the water overnight, the usual practice is to tie the boat up either in the shelter of the inside bow or under the inside quarter (that is, the side nearest to the wharf), according to the direction of the wind or tide.

To make fast alongside a stone wharf or one constructed with piles is simply asking for trouble. Not only is the boat likely to be damaged through bumping the wharf all night, but it is also liable to be trapped under horizontal piles with a rising tide.

The safest thing to do is to make fast with one line on the wharf and another on board, in such a position that she will be clear of any soil pipes. Sufficient allowance must be made for rise and fall of the tide on the line connecting with the wharf, and when all the gear (if any) has been removed, the line on board is tightened up. This will allow the boat to ride comfortably between the ship and the wharf without bumping.

Temporary Painter.—If a lifeboat is to be used in harbour or other sheltered waters, a length of small size line should be put in the boat for use as a temporary painter. Those fitted as part of the equipment are very unhandy in quiet waters.

Unable to Reach Ship.—Should it happen that you are unable to reach your ship on account of strong tides or wind, make for a position astern of her, and those on board might be able to float or veer a line down to you.

Care of Boats.—Whenever there is time, wooden boats are lowered into the water and left there as long as possible. This is done to tighten up the seams which, through becoming dry, are liable to shrink and open up. A period in the water will cause the wood to swell and close the seams. Very often in tropical weather a few inches of water is put in each boat (with a hose) for the same purpose.

Boat covers are also spread to keep the sun's rays off the planking, but these are usually removed when nearing crowded waters—especially on large passenger vessels.

MISCELLANEOUS BOATWORK.

Gripes.—A boat hanging at the davits would swing a good deal with the movement of the ship if it were not lashed in position.

For this purpose, wire gripes (covered with canvas) are fitted to each davit head, and these are long enough to cross on the outside of the boat and reach a slip arrangement on the deck.

To these they are made fast with lanyards after heaving tight with a "handy billy". The gripes cover a large area of the boat, which it holds tightly against the griping spar.

Griping Spar.—A long, slightly tapered spar which hangs between, and outside radial davits. Two corked-filled canvas fenders are fitted to the spar, and these act as a cushion when the boat's side is hove tightly against it.

A small tackle on each davit holds the weight of the spar at a height of about 3 feet from the deck, and the spar is also lashed to the davit with a cross lashing.

When about to enter port and the accident boat has to be swung inboard, the spar is lowered down on the deck.

Lazy Painter.—The extra painter fitted to the thwart of an accident boat by means of a toggle. It is easily slipped, and no time is lost in getting away from a vessel's side through hauling in the painter.

Tillers.—Accident boats are often lowered while a vessel is still moving through the water. When this is the case, it will be necessary to steer the boat as soon as waterborne to keep her close to the ship's side while the falls are disengaged. After that she must be kept away from the side, especially if the blocks are swinging about.

In cases of this kind it will often be found that the after fall interferes with the free action of the tiller. This is liable to lead to an accident, and the only remedy is to provide a short tiller which will allow of free movement while the fall is hooked.

Boarding Wrecks.—Sometimes a much smoother passage can be made for an accident boat by pouring oil overboard to windward of a wreck and then taking advantage of the lee created by both vessels.

Never approach too close to a wreck on the weather side. The rebound or backwash of the sea from her side may be sufficient to upset the boat.

Should there be danger from falling spars or other wreckage on the lee side, the bow or stern can usually be approached with comparative safety, but to prevent contact with the wreck, it might be necessary for the wrecked seamen to jump into the water and be pulled into the lifeboat.

Picking Up a Wrecked Crew.—With a lifeboat full of wrecked seamen safely alongside the rescue ship, it will be found easier for them to come on board if cargo nets are hung over the side, in addition to the usual pilot ladders.

As they cover a large area, a foothold may be gained irrespective of the position of the boat, and a number of men may climb the side without loss of time.

Carrying Out a Stream Anchor.—A small stream anchor may be hung under the after end of a lifeboat by means of a good lashing passed over both gunwales, just forward of the stern-post, and if necessary, a short piece of timber can be laid across to take the weight of the lashing off the gunwale. The mooring warp attached to the anchor may be either paid out from the ship, or the whole, or part of it, carried in the boat.

As the shackle on the runner will probably be under water when the boat takes the weight, it will be necessary to attach the runner to the anchor with a long strop in order that the shackle will be within reach above water. Fig. 10.

If an anchor is too heavy to hang under the stern of a boat it may be slung amidships, if a good, stout plank, long enough to project at each side, is lashed firmly across both gunwales.

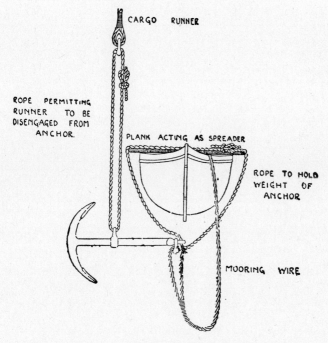

CARGO RUNNER

ROPE PERMITTING
RUNNER TO BE
DISENGAGED FROM
ANCHOR.

PLANK ACTING AS SPREADER

ROPE TO HOLD
WEIGHT OF
ANCHOR

MOORING WIRE

FIG. 10.—Carrying Out a Stream Anchor.

When the shackle of the anchor is lowered to the water's edge, a suitable lashing is passed through it and up both sides of the boat to form a bridle. Lower away until the boat takes the weight, and the anchor will hang directly under the boat. If a strop has been used for connecting the runner to the anchor, it can easily be let go from the boat.

Carrying Out a Bower Anchor.—Two boats and two good spars are required to carry away a really heavy anchor, and a certain amount of judgment is necessary when spacing the spars, if the boats are to sit properly in the water with the load. Lower the anchor into the water with the shank horizontal, flukes vertical, and place a boat on each side of it. Lash the two spars securely across both boats, one directly over the flukes, and the other above the shackle. When the spars are lashed, pass a good lashing under the top fluke and round the forward spar, and also pass another lashing under the shank and round the after spar. When all is fast, lower away easy and if the spars have been properly spaced, the boats will sit evenly, or a little by the stern.

Good judgment is required when cutting adrift in order that both lashings will part at the same time.

Lashing Boats.—When large quantities of goods have to be transported by water, it is sometimes the practice to lash two boats together, deck them across, and form what is practically a big raft. These are known as lash boats, and large loads can be carried, compared to the comparatively small cargoes of a single boat. Lash the two boats together with a good bow and stern lashing, and lay a good spar across the ends of the boats, about 3 feet from the bow and stern-posts.

When these are lashed down, lay fore-and-afters in such a way that there is one parallel with, but a little to the inside of both outboard gunwales, and one on the inside of each inboard gunwale. The whole framework must be lashed together and down to the boat before the planks are laid across. The number will depend on the length of the boats and the size of the planks. These are usually about 9 × 2 in., while the heavier spars are about 4 or 5 inches square. The planks are secured with cross lashings, and provision should be made for access to the bottom of the boats for bailing out.

Using an Oar.—Nothing stamps the amateur so much as the habit of describing a large arc with the blade of an oar. Any tendency on the part of an oarsman to do this should be instantly corrected.

Stroke Oar.—It is always advisable to start off with a very slow stroke when a boat's crew is made up of firemen and stewards. Chances to practise are few and far between, and they may be some time getting used to the oar again.

Sculling With Single Oars.—This is a very handy means of propulsion when only one man is in a boat. To a sailor in a coasting vessel it is as necessary as being able to steer. The blade is made to describe an arc under water with a side-to-side movement of the hands, and the weight of water holds the oar down on the row-lock or groove cut in the transom in the case of a jolly-boat. Actually it may be likened to a quick method of levering water astern, thus causing the boat to move ahead. Practice is essential to acquire the art, and although lifeboats are not fitted for the purpose, a certain amount of practice could be done with an oar at the side of a boat.

DEFINITION OF ROWING ORDERS.

Up Oars.—The oars are raised smartly into the air and held in a vertical position, blades fore-and-aft with the looms resting on the bottom boards.

Out Oars.—To put the oars into the "oars" position ready for pulling.

Give Way Together.—The signal to begin pulling.

Oars.—The signal to stop pulling. After hearing this order the crew take one more stroke, then stop with the oars parallel to the water, at right angles to the fore-and-aft line and with the blades horizontal.

Rest On Your Oars.—This order is given when it is intended to lay-to for a while, and is the equivalent of the soldier's "stand easy". The oars are pulled in a few inches so that they are easy to balance. This order very often comes after "Oars".

Back Water.—To work the oars in the opposite direction to that taken when pulling. This order is only given when a boat has to be stopped quickly, or when manoeuvring.

Back Port (or Starboard).—Only the oars on the side indicated would back water when the order is given. It is used when turning a boat short round or when manoeuvring near a landing place.

Back Together.—To "back water" with all port and starboard oars.

Bow.—In order that the bow oarsman will have sufficient time to stand by with a boathook, he is always ordered to take his oar in before the others.

Way Enough.—This order indicates that the boat has enough speed to reach her objective. The crew take one more stroke and should then boat the oars. The orders "Toss and Boat the Oars" "Boat the Oars", or simply "In Oars", may or may not be given.

Toss and Boat the Oars.—To toss the blade of the oar smartly in the air, then lay it in the boat, blade forward.

In Oars.—To pull the oar in and lay it in the boat, blade forward. In a "whaler" the blade would be aft.

Orders Under Oars.—After giving the order "Let go the painter", it is better to hang on to the side ladder for a few moments, until the painter is hauled in by the bowman.

In the interval, see that the crutches are shipped, and when ready, give the order "Shove off". (This is done by the bowman and stroke oar.)

With a well-trained crew it would be possible to give the order "Oars up" before getting under way, but this is seldom done in the Merchant Service. "Out oars" is usually the term used, and this is followed by "Give way together", when everyone is settled and ready to start pulling.

The only orders likely to be given while under way are: "Oars", which may be followed by "Rest on your oars", "Back together", "Back port", or "Back starboard".

When about to go alongside a landing-place, the coxswain should give the order "Bow" a few moments before the speed of the boat warrants "Way enough". This may be followed immediately by "Toss and boat the oars", "Boat the oars", or simply "In oars".

Steering Orders.—When steering it is only necessary to remember that the order "Hard a starboard" or "Hard a port" refers to the rudder and not the tiller. If the order is "Hard a starboard", put the tiller over to **port** and the boat's head will come round to **starboard**. If the order is "Hard a port", put the tiller to **starboard** and the boat's head will come round to **port.**

Turning Short Round.—To turn short round to starboard, give the order "Back starboard—Pull port", and *vice versa* if turning to port.

SAILS AND SAILING.

Stepping the Mast.—If the boat is jumping about, use the stays for steadying the mast as it goes up, and make sure the sheaves lie in a fore-and-after direction before clamping it in place.

In smooth water the usual practice is to loosen and clear the stays and halliards after the mast is stepped, but it is as well to make sure beforehand that everything is actually clear.

Sail Types.—The only sail used for lifeboats in the Merchant Navy is the Standing Lug but if the boat is 25 feet or over in length then a Jib must also be carried. (See Fig. 11.)

These sails are made of the best quality duck and are fitted with reef cringles and reef points. The roping, is sewn round the head, luff, tack, clew and up the after-leech as far as the reef cringles. The jib is only roped on the luff.

Bending Sail on Yard.—Lug-sails are bent on the yard with throat and peak lashings, and although the head of the sail is then nearly always laced with boat-lacing from the centre towards each end, a separate stop at each eyelet hole is better. This prevents the head of the sail coming adrift if the lacing should be chafed through.

The roping is always on the port side of the sail when set.

Hoisting and Setting Sail.—Lay yard and sail fore-and-aft on what is to be the lee side. Hook tack on "horse" if a dipping lug and take a loose turn at foot of mast if a standing lug. Stretch sheet aft, hook yard on traveller, order all hands to windward, then hoist away. Set up good and tight on the halliards, then heave down on the tack until you see a crease running from the peak towards the tack. When full of wind the crease will disappear and leave a well-set sail. A badly-set sail interferes with the sailing qualities of a boat and is very unseamanlike. One common cause for this is the strop on the yard getting out of position. As a general rule it should be one-third the length of the yard from forward for a dipping lug, and one-quarter for a standing lug.

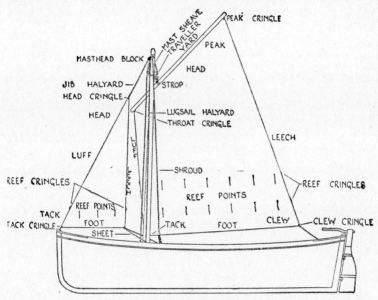

FIG. 11.—Standing Lug.

Another point to watch is that the yard is not less than 12 inches below the lug-sail halliard sheave, and the tack is not more than about a foot above the thwart. If the lug-sail halliards are made fast on the weather side, and a little forward of the mast, they will give extra support in heavy weather. Also, when the jib halliards are set up tight, the roping on the jib, along with the halliards, will form a good fore-stay when diving into a head sea.

Standing Up in Boats.—There is no necessity to stand up in a boat—especially when sailing. It is very bad form, and conveys the impression of a crowd of holiday-makers at the seaside. Practically everything can be done sitting down.

Sailing Terms.—A boat cannot sail directly into the wind's eye but makes a zig-zag course when sailing against the wind. This is known as **Tacking.** When the wind is on the port side a boat is on the **port tack,** and with wind on the starboard side she is on the **starboard tack.**

When sailing as close to the wind as possible (luff shaking), or as long as the wind is anywhere on the bow, a boat would be **close-**

G

hauled, and according to which bow, may be either **close-hauled on the port tack** or **close-hauled on the starboard tack.**

"**Ready about**" is an order given as a warning that the coxswain is preparing "**To go about**", which is the manoeuvre of changing from one tack to another. To do this keep **Clean full for stays** or, in other words, let the boat run off the wind a little until the sails are full, so that she will gain speed.

When ready, **down helm** (tiller to leeward), and **let go jib sheet** to ease the pressure forward. As the boat comes round **flatten aft the lug sheet** (pull tight), to increase the pressure aft and assist the rudder. When head to wind with the sails shaking she will be **in stays.** At this point "**Back the jib**" to help her round, and when boat is well round let go jib, **flatten sheets** and steady up for new tack. If the boat's head had fallen off again instead of coming round on the new tack, it would be said that she had **missed stays.**

If the weather is bad and tacking is impracticable, **Wearing** may be resorted to, but a lot of ground is lost in this manoeuvre as the boat has to describe an almost complete circle which will take her away from the direction in which the coxswain wishes to proceed. When ready, put the **helm up** (tiller to windward), **ease off lug sheet,** and when wind is right aft, **let go jib sheet** and take a loose turn on opposite cleat.

When the boat is round with the wind on the new weather side, **flatten aft lug sheet, square away** on new tack, and **trim jib sheet** as she comes up into the wind. If there is much wind it is safer to lower the sail down until boat comes round, then hoist again.

A boat with the wind **abaft the beam** is **sailing free.** When altering course the wind is brought from one quarter to the other by **gybing,** in exactly the same way as for wearing.

To relieve the pressure on the sails when close-hauled during a squall, **luff her through it** by putting her head up in the wind until the sails are shaking. If sailing free, **keep her away,** but if the squall is stiff the sail is best lowered down altogether.

Sailing Hints.—When sailing with the wind right aft the sheet should be eased off until the sail is nearly at right angles to the keel. As the wind draws out, haul in the sheet until the best results are obtained. When sailing close-hauled the sheet should be nearly flat aft.

It is better to run with the wind first on one quarter and then on the other, instead of dead aft. This avoids the risk of accidentally gybing, the jib draws better, and the boat sails better.

A heavily laden boat should be either well reefed down in a strong breeze or under oars. The danger of capsizing is small, but she will not rise buoyantly, and may drive through or under a sea and fill up.

The position for the crew when sailing is abaft the mast on the lower thwarts. Never allow anyone to stand on the thwarts or climb the mast.

The coxswain sits on the weather side, the sheets are turned up on the lee side, but are never made fast. They are held in the hand, ready for slacking away quickly.

Never jam a tiller over suddenly. This stops the way on the boat. Put it over gradually, and then only a little more than half way.

Making-up Boat Sails.—Never roll a sail around a yard like a window blind. Fold in the tack and sheets and finish off by rolling the sail towards the yard, to which it should be lashed tightly with one turn of rope-yarn at about four equi-distant places.

If it is made up lumpy or not lashed securely, trouble will be experienced in sliding it into the cover. The rope-yarns should be tied bow-fashion, so that they may be easily cast adrift if a knife is not available. Always bear in mind that the sail has to be used again, therefore it should be left so that no trouble will be experienced in getting ready for hoisting next time it is required. It is a good plan to leave a short length of tack and sheet hanging out the respective ends, so that they may be quickly found and stretched to their places, without having to search for them.

Stowing Mast and Gear.—When a mast is not in use the usual practice is to stretch the lug and jib halliards along the mast and make them fast in that position.

It is better not to cross the backstays, but to simply stretch them like the halliards, then pass either one or two light lashings with a rope-yarn, at a height convenient for cutting from the thwart. Make the gear fast in the order in which it will come off, keeping each part separate, for simplicity when letting go.

Wedge for Mast.—Should there be any play between the mast and the mast clamp a wedge should be used to prevent any movement. Bore a hole in it so that it may be attached to the thwart by means of a lanyard, to prevent loss.

Rope-yarns in Boat.—A few strands of rope-yarns in a lifeboat may prove of great value on a number of possible occasions. These may be taken from any old rope which happens to be in the boat, but to be on the safe side, surplus rope should be cut from the life-lines before leaving the ship's side.

Bailing Out Boats.—Some motor launches are not fitted with a draining hole and water lying in the bottom has to be pumped out— very often a long job after heavy rain. A quicker method is to employ the syphon principle by utilising a rubber hose. Connect up to a hydrant in the ordinary way and immerse the opposite end in the water to be bailed out. Open valve of hydrant, and when water is running strongly shut off and disconnect quickly and the water will flow from the boat, provided the hydrant end is beneath the level of the water in the boat.

Warped Oars.—As a rule when oars are laid in a boat they only rest at the ends (loom and blade), the middle part (shaft), sagging, through the sheer of the boat. In time the oars will become warped very often so badly that they will be useless for rowing.

To prevent this, either one or two chocks should be fitted to the side benches, in such a position that they will support the middle part of the oars.

Marking Mast.—To save time when stepping a mast it is a good plan to cut a small F (indicating forward) on the forepart of the mast.

For the benefit of men with little experience in boat work this idea is very useful, particularly at night.

CHAPTER VII.

CLEANING GEAR

Cleaning Materials. Washing and Scrubbing Decks. Lamps, Lights, Oils. Tools. Various Deck Stores.

CLEANING MATERIALS

Soda.—Large quantities of ordinary washing soda are used on all vessels, particularly for washing paintwork.

Sugi Mugi.—All solutions used for washing paintwork are generally termed sugi mugi, although there is a special brand of soap powder which bears that name.

Soft Soap.—The two principal constituents are oil and potash. Various oils are used in its manufacture, such as whale or seal-oil, and sometimes linseed oil, fish oil, or tallow. Besides being a good cleaning agent, soft soap is invariably used in shipyards for greasing the launching ways of ships.

Caustic Soda.—A powerful alkali used for cleaning purposes. It will quickly cause a burn if it comes in contact with the skin.

Special care should be taken to protect the eyes, and for this reason salt water, which will quickly kill it, should always be kept handy when caustic is in use.

Atlas.—A strong cleaning agent used for the same purpose as caustic soda.

Climax.—Similar product to Atlas. Is used for all purposes where a strong solution is required.

Wads.—A bundle of waste, a piece of mutton cloth, or any old rag for that matter, is known as a wad, and may be used for washing paintwork, mopping up water, polishing brass, etc.

Mutton Cloth.—These are the old (sometimes new) cotton cloths which cover frozen meat carcases. They are washed before being sent on board, so are entirely free from grease.

Waste.—Refuse material from textile factories. Its uses are well known, and it is used extensively on board ship.

Of the two classes—white and coloured—the former is the cleanest and best.

Tank Flannel.—Thick absorbent material used for drying surfaces when cleaning tanks.

Sponge Cloths.—A material which is similar in appearance to mutton cloth and which is used for the same purpose.

Hoses.—Rubber hoses are often used for washing down decks. These are distinct from the fire hoses, which are kept specially for that purpose.

Deck Brooms.—Long-handled brooms with long bristles. The same type is used on all vessels. Fig. 1.

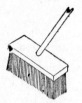

FIG. 1.—Deck Broom.

FIG. 2.—Turk's Head.

Turk's Heads.—Long-handled soft brushes used for washing paintwork. Fig. 2.

Note.—This is also the name of an ornamental knot which derives its name from the supposed resemblance to a turban.

Paint Scrubbers.—A wide, soft-haired, short-bristled brush used for scrubbing paintwork. Fig. 3.

FIG. 3.—Paint Scrubber.

FIG. 4.—Bear.

Bears.—Two types of bears are in general use, one with ordinary stiff bristles, and the other with steel. The iron frame weighs about 40 lbs. and the brushes, which are fitted into a slot arrangement, can be replaced when worn out. Fig. 4.

Holystone.—A heavy stone resembling sandstone. It is encased in a light frame attached to a long handle and used for scouring wooden decks. Another purpose served by the holystone is to remove " dead " wood from the surface of the deck. Fig. 5.

Mufflestone.—A holystone with an ordinary door-mat tied to the underneath side, with bristles to the deck.

Squee Gee.—To dispose of surplus water on a wooden deck, and to help in the drying process, a long handled frame containing a strip of rubber is pushed along the wet surface. The rubber clings closely to the wood and pushes water in front of it. Fig. 6.

Mop.—In addition to the purpose for which they are intended, mops are often used for oiling iron decks, or wetting wooden decks with " strongers " (a strong cleaning agent), before " holystoning " or " barbarising."

Brass Cleaning Tin.—A shallow tin box with a hinged lid and handle. It is divided into compartments which hold a mixture of colza oil and pulverised bathbrick.

Metal Polish.—The ordinary types of metal polish will be found on board many vessels.

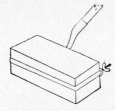

Fig. 5.—Holystone.   Fig. 6.—Squee Gee.

Bathbrick.—A compound of sand and slime deposited by the River Parret at Bridgwater, which is the only place where it has ever been found. It is dug up at low-water and sold in the shape of small bricks. Is used extensively for cleaning brass and paintwork.

Pumice Stone.—A light stone of volcanic substance, imported mostly from the Lipari Islands, and used for polishing or smoothing rough surfaces.

Emery-Paper.—The function of emery-paper is well known and the grades are as follows: O, F.F. (Double F), F, 1, 1½, 2, 2½.

Sandpaper.—The grades are as follows: 0.0 (very fine, almost like flour); 0, 1, 1½; F. 2 (Fine 2); M. 2 (Medium 2); S. 2 (Strong 2); 2½, 3. It is sometimes called glass-paper, which is more correct, as crushed glass is employed as the abrading material. Originally fine sand was used—hence the name.

WASHING AND SCRUBBING DECKS.

Holystoning.—The work of keeping a ship clean at sea is one continual round of washing paintwork and holystoning decks. The ordinary process of " washing down " with a hose and deck brooms is all right for the iron decks of tramp steamers, but in passenger ships the wooden decks require something more effective to keep them white and clean. For this purpose holystones and bears are mostly used. After wetting the deck with fresh water and sprinkling sand over it, the holystone is worked to and fro with the grain of the wood. Bears are also used in the same way, and being much heavier they are more effective, especially the steel ones.

To help this process, a strong solution of either caustic soda, Atlas or Climax is spread on the deck first, but if made too strong it is liable to burn the wood and give the deck a brownish tinge.

Barbarising.—After the deck has been wet over with caustic, Atlas or Climax, and sand has been laid down, the decks are vigorously scrubbed with deck brooms.

Handstoning.—In odd corners and close to bulkheads, where a holystone or bear cannot reach, a handstone is used in the same way as the larger ones.

Sand and Canvas.—Woodwork kept bare of paint or varnish is cleaned occasionally with sand and canvas. Wet canvas with a sprinkling of sand is rubbed on the wet woodwork until clean.

Scrubbing Machines.—Electric scrubbing machines are usually found in passenger ships which have wooden decks.

Removing Rusty Spots.—If ugly yellow rust marks cannot be removed with strong sugi, they should be rubbed with sand or bathbrick, or anything of a gritty nature.

LAMPS, LIGHTS, OILS.

Details of the Navigation Lights, i.e. the area over which the lights show, the heights at which they are carried and the ranges are given in the "Regulations for Preventing Collisions at Sea", commencing on page 281.

Navigation or Steaming Lights.—Most ships carry electric navigation lights which are fixed permanently in their respective positions, and in addition, must carry oil lamps or battery powered secondary lights.

Masthead Lights.—As the name implies, these are hoisted on the foremast and mainmast respectively, and show a white light. In the case of oil masthead lights, the lantern is secured in a special cage or carrier which runs up on two wire stays on to a rigid fork whose purpose is to hold the cage and hence the light steady in position. The cage is hoisted by means of halliards.

Sidelights.—The port light (*red*) and the starboard light (*green*) are fitted into special screens at each side of the navigation bridge. The screens are painted matt black.

Stern Light.—A white light fitted into a special screen at the after end of a vessel.

Anchor Lights.—When a vessel is lying at anchor two lights must be exhibited. One is hoisted on the lamp halliards attached to the forestay, and the other on the flagstaff aft. Both are white, and must be visible all round the horizon. They may be either oil or electric lamps.

Not Under Command Lights.—When a vessel is out of control she must exhibit two red lights which are visible all round the horizon. A contrivance holding the required lights is kept handy on the bridge, and they are hoisted on the jumper stay from there. Two red oil lights are also to be found in the lamp locker.

Binnacle Lights.—Each binnacle is provided with at least one oil lamp in addition to electric light, and this is kept shipped in its proper place and ready for use.

Hurricane Lamps.—These are a well-known type of storm lamp, and are used for all purposes where a light is required in the general work of the ship. Fig. 7.

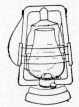

Fig. 7.—Hurricane Lamp.   Fig. 8.—Duck Lamp.

Clusters (or Portable Cargo Lanterns).—A cluster is an enamelled reflector (about 18 inches diameter), containing a number of electric light bulbs, and attached to a long wire lead. It can be connected to special switch boxes, situated at strategic points around the deck. They are mostly used down the holds when working cargo at night.

Floodlights.—Special lights fitted for the purpose of "flooding" certain areas with light, either for ornamental effect in the case of funnels, or to facilitate the handling of cargo when placed near the ship's side.

Duck Lamps.—Small containers holding colza oil, with a spout from which the end of a coarse long wick protrudes. They are naked lights, and were only used by coal trimmers in the bunkers. Fig. 8.

Accident Boat Light.—This may be either a hurricane lamp or the type of storm lamp usually found in lifeboats. It is usually lighted and put in the accident boat each night, so as to be ready for use if wanted.

Emergency Lights.—A lighting system which extends over the whole ship and which comes on automatically in the event of a power failure.

It is distinct from the normal locally controlled lighting system, and is for use at such times when normal lighting is switched off, so that people may find their way about decks and alleyways.

Colza Oil.—An oil found on practically all vessels. It is made from crushed rape-seed and used as a lubricant, as well as an illuminant.

Paraffin Oil.—Is mostly obtained by distillation from crude petroleum, and is used extensively as an illuminant.

Turpentine.—A colourless oil of pungent odour, obtained by distillation from the resin of various coniferous trees. Large quantities of this oil come from the forests of North Carolina. It is used extensively in the mixing of paints and varnishes. It also contains medicinal properties.

Linseed Oil.—Is made from an oil and nitrogenous matter contained in the seed of the flax plant. The residue is made into cattle food.

Boiled Linseed Oil.—The boiling of linseed oil takes about five hours, and during this time driers are added in the proportion of about 5 lbs. of driers to 1 ton of oil.

Fish Oil.—This may be derived from all parts of a fish, but oil from livers is considered best where large fish are concerned. Oil from sardines caught on the Spanish coast is said to be best of all.

All types of fish may be used in the manufacture of fish oil, such as sharks, salmon, cod, sprats, etc. Fish oil is sometimes blended with cheaper oil for use on decks, wires, etc.

Oiling Wooden Decks.—Wooden decks are oiled for the purpose of preserving the wood. Raw linseed oil is used for the purpose, and it should be applied with a piece of soft material such as flannel or mutton cloth. Waste leaves unsightly bits of cotton adhering to the rough surface and is seldom used. The deck is always well barbarised or holystoned first, and if a little red-lead is mixed with the oil in the proportion of $\frac{1}{2}$ lb. to 1 cwt. of oil, it acts as driers, and a more pleasing effect is obtained. To ensure that the deck dries quickly, the oil should be put on as thinly as possible.

If a little fish-oil is added to the raw oil it will give an effect similar to varnish. Teak decks are never oiled. Raw oil has the effect of turning them black.

Oiling Steel Decks.—Many methods of preserving steel decks are employed, but none are entirely successful. The most popular method appears to be the application of fish-oil. This is laid on with either wads, turks' heads or mops. Fuel oil is also used extensively, and has the advantage of being able to loosen rust after a few applications. Quick-drying, boot-topping composition paint is also used in some ships, but this practice is dying out fast. All loose rust must be scraped off and the decks well swept before anything is put on.

Oiling Wires.—Wires that are not painted, and mooring-wires in particular, are frequently treated with a coat of oil to prevent them turning rusty. Fish-oil is used, and is applied with a brush. When a wire has already turned rusty, a small hand wire broom is used to scour it first, especially in the case of a mooring-wire with accumulations of mud and old coats of oil adhering to it. White-lead and tallow is another good preservative for wire, but it is only used on standing rigging.

Mineral Sperm.—Sperm oil from whales mixed with a hydrocarbon oil which has a high flash-point. It is considered the ideal oil for navigation lights, but is rather expensive.

Stockholm Tar.—A softwood tar which is variously known as Pine Oil, Swedish Tar, Finland Tar, or Russian Tar, according to the country from which it is exported. It is obtained by carbonising the resinous root-stumps of various conifers, especially old pine roots. Stockholm tar is a good preservative and has excellent waterproofing qualities.

Coal-tar.—A fluid extract derived from coal when making gas.

Tallow.—That generally in use is a blend of real tallow and animal fats from an abbatoir. The three grades are known as No. 1, No. 2, and Blended.

Blacklead.—This is another name for graphite, which is also known as plumbago. It is used as a lubricant on surfaces where friction occurs and oil is undesirable. On board ship it is mixed with tallow and used for greasing the pins and sheaves of blocks, gins, etc.

Log Oil.—A specially refined oil to ensure the smooth working of the patent log.

Engine Oil.—Heavy oil kept for lubricating machinery such as winches, and movable parts like warping chocks, etc.

Barrels.—The average barrel of oil contains about 40–50 gallons of oil and weighs about 450 lbs. The ends of a barrel are each known as the "head". Starting from each "head" the iron bands, or hoops, are known as the "chime" (usually two), "quarter" and "bilge". As the rivets on the hoops are in line with the bung, its whereabouts may be determined without any trouble in the dark.

Emptying Barrels of Oil.—Barrels of oil belonging to the deck stores are emptied into tanks in the lamp locker. A special brass tap can usually be found for this purpose, and this is hammered into a hole in the head of the cask. The hole is bored with an auger about one inch in diameter, and should be bored on the opposite part of the circle to the bung, so that when the barrel is laid on its bilge the tap will be at the bottom, and the bung, which is used as an air vent, will be uppermost. The oil will run very slowly from the tap unless an air vent is introduced, and to save the trouble of extracting the bung, a small hole is bored in either the head of the cask or in the bung itself. The barrel must be built up high enough to allow a bucket to be placed under the tap; if a tap cannot be found, an ordinary wooden plug must be used, in which case the air vent should be kept nearly closed if the job is to be performed without waste. The tap must be fitted before the cask is laid on its bilge.

Oil is usually carried in steel drums of 40 gallons, and in one head there are two hexagonal plugs, one to enable a tap to be fitted and the other to act as a vent. A common practice is to use an air-operated pump with a suction pipe long enough to reach the bottom when the drum is standing on its head.

TOOLS.

Blowlamp.—Although it may not be found as part of the deck stores, the ordinary type of blowlamp used for burning off old paint can normally be borrowed from the engine-room. Today, paint stripping is often done by the use of a special stripping fluid which softens the old paint so that it can be scraped off.

Chipping Hammer.—Small hammer weighing about ½ lb., with two sharp edges set at right angles to each other. It is used for chipping thick rust or paint from ironwork.

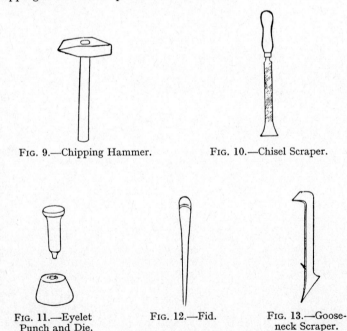

FIG. 9.—Chipping Hammer.   FIG. 10.—Chisel Scraper.

FIG. 11.—Eyelet Punch and Die.   FIG. 12.—Fid.   FIG. 13.—Goose-neck Scraper.

Chisel Scraper.—These are about 2 feet in length, and have a broad edge. They are used mostly for smoothing off rough surfaces.

Cold Chisel.—A specially hardened piece of steel with a good cutting edge. It is used mostly for cutting wire strands when finishing off a splice.

Cold Set.—Large-sized cold chisel with a handle attached. This permits heavy blows to be struck without risk of injury to the hands. Cutting wire rope or rusted nuts is practically the only use to which it is put.

Crow-bar.—Strong steel bar used for all purposes where extra strong leverage is required.

Eyelet Punch.—The tools used for closing brass eyelets are a punch and a shaped brass iron known as a "die". Two punches will usually be found on board—a small one and a large one. Each can be used for about four different sizes.

Fids.—Tapered and pointed pieces of lignum-vitae used for splicing ropes. They are similar to a marline-spike, but are much thicker at both ends, and are made in a variety of sizes.

Another type of fid is a substantial iron bar which fits through the lowermast, for the purpose of supporting the heel of the topmast.

Gooseneck Scrapers.—An excellent tool for scraping pitch from a deck. It is simply a broad piece of metal about 12 inches long with the end bent over at right angles to the handle and sharpened.

Handspike.—A long piece of wood closely resembling a capstan bar, and used for any job where leverage is required. Most "first trippers" confuse it with a marline-spike on account of the name.

Long Handled Scraper (or Slice).—A wide-bladed scraper with a pole handle. It is used for scraping loose rust on iron decks.

FIG. 14.—Handspike.   FIG. 15.—Marline-spike.

Marline-spike.—Tapered and pointed steel bars used when wire-splicing. With these the strands of a wire are prised up. They are usually called "spikes" for short, and are made in various sizes, ranging from a few inches to 2 or 3 feet in length. An old one is always kept for slackening off shackle pins and other rough work.

A spike with a flat chisel point is considered to be the best for wire-splicing. Another good type has a groove near the point to provide an easy passage for the wire strands.

Mast Knife.—These are about 10 inches long and closely resemble a sheath-knife. They are used for all scraping purposes, but never for cutting.

Munday Hammer.—An exceptionally heavy hammer which is only used when really heavy blows are required. They are not always carried among deck stores but in any case can be borrowed from the engine-room.

Oilcan.—The ordinary type with plunger feed and holding lubricating oil will be found among all deck stores.

Oil Feeder.—Large type of oilcan, more after the teapot style. In the engine-room they mostly hold lubricating oil, but on deck are kept exclusively for paraffin.

Palm.—Leather guard which fits over the palm of the hand when sewing canvas. A small indented "iron" is fitted in the right position for pushing on the end of the sail needle. Fig. 16.

Fig. 16.—Palm.

Fig. 17.—Sailhook.

Pinch Bar.—Small, handy type of crow bar.

Pipe Wrench (or Footprints).—A tool by means of which round objects (pipe and round nuts) are turned or unscrewed. The grip is effected with sharp teeth and a scissor movement leverage. They are often called "footprints" because that was the original makers trade mark.

Podger.—Small iron bar with pointed ends which are bent to opposing angles. They are used specially for tightening up bottle screws. Fig. 18.

Fig. 18.
Podger.

Fig. 19.
Rat Tail Spanner.

Rat Tail.—These are only found on comparatively new vessels, and may have been left by shipyard workmen. Having a tapered end they make handy marline-spikes, although they are really designed as spanners. Fig. 19.

Rigging Screw.—When a wire has to be bent almost double to take the shape of a thimble before splicing, a rigging screw is used to both hold and screw the wire into position in the same manner as a vice. Fig. 20.

Rope Gauge.—A handy instrument which is used for measuring the circumference and the diameter of wire, rope, or chain. Fig. 21.

Sail Needles.—A triangular bladed needle used for sewing canvas. They are made in various sizes and numbered from 5 to 16.

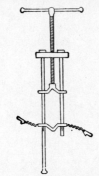

Fig. 20.—Rigging Screw.

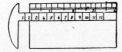

Fig. 21.—Rope Gauge.

Serving Board.—When serving a rope or wire, marline is bound very tightly around the part to be served, and to get it properly tight a serving board is used, or a serving mallet. Figs. 22, 23.

Fig. 22.—Serving Board.

Fig. 23.—Serving Mallet.

Shackle Key.—A T-shaped tool with one of the ends finished off square, and the other two adapted for use as a screwdriver. Fig. 24. It must be used in cases where a shackle-pin has a square hole in the end, or when it is slotted in the same manner as a Screw.

FIG. 24.—Shackle Key.

Shifting Spanner.—Spanners used on shipboard are exactly the same as those found ashore.

Sounding Rod.—A metal rod, either in one length or jointed to go round bends, and graduated in inches or centimetres. It is lowered down the sounding pipes by a light line to take measurements of the water in the bilges or in a water tank.

Sounding Tape.—A graduated steel tape with a brass weight. It is used to measure the depth of oil in oil-tanks.

Ullage Tape.—A tape used to measure the space above the oil in a cargo tank.

Three Cornered Scrapers.—Both small and large ones will be found on board. While the former are very handy and useful when scraping varnish or paint, the latter are of no practical use, except for scraping dirt from the beds of winches, etc. Fig. 25.

FIG. 25.—Three Corner Scraper.

FIG. 26.—Topping Maul.

Topping Maul.—A special type of hammer used by all ships' carpenters. They have a wide flat head and a blunt, tapered point. Fig. 26.

Wheel Spanner.—Short lever with branching ends adapted to fit the spokes and rim of a valve wheel.

VARIOUS DECK STORES.

Anchor Balls (or Shapes).—A ball consists of two round metal sheets about 2 feet in diameter. These are set at right angles to one another, and in most cases they are collapsible for stowing away when not in use. It is always painted black. Fig. 27.

Fig. 27.—Anchor Ball.

Apron.—The canvas screen against which a leadsman leans when heaving the lead, and which protects his legs from the wet line.

A canvas apron of the usual type is also used by the man holding the hose when washing down.

Arming.—Mixture of tallow and soap inserted in a special cavity in the bottom of all sounding leads. This is done for the purpose of picking up particles from the sea bed, so that they may be compared with the information given on the chart.

Ash Bags.—Special strong bags used by firemen when ashes have to be hove up from the stokehold and dumped overboard.

Ash Buckets.—Heavy iron buckets used for the same purpose as ash bags. They are always kept in racks in the stokehold entrance.

Awnings.—Canvas covers spread above the deck to minimise the effect of strong sunlight on the living quarters, and also to protect the deck planking. They lie over awning spars bolted to awning stanchions through which a ridge wire is rove. The awning is made fast to the ridge wire with "stops", which are just short ends of rope attached to the awning itself.

Bannister Brush.—These are only found on vessels carrying frozen or chilled meat, and are only used when cargo space has to be carefully cleaned out before loading.

Fig. 28.—Beam Clamp.

Beam Clamps.—Special clamps with a ring attached. These can be fitted to any of the frames or beams within the hold or elsewhere. They permit a snatch block to be used in out-of-the-way corners for convenience in moving heavy weights. Fig. 28.

Boat Rope.—Practically any piece of line of sufficient length and size could be used as a boat rope. One end is made fast well forward on the foredeck, and the loose end thrown to any boat which wishes to come alongside.

It is always kept near the pilot ladder, so as to be handy when wanted. A 3-inch rope is very often kept specially for this purpose.

Bo'sun's Chair.—A flat piece of wood about 18 ins. by 5 ins. by 1 in. with two holes at each end through which a strop is rove and spliced. Fig. 29. (*See also* page 80.)

FIG. 29.—Bo'sun's Chair.

Bottle Screw.—This contrivance is used mostly for "setting up" stays, shrouds, ridge wires, steering chains, etc.

It is an easy method of hauling things tight and keeping them in that position. Fig. 30.

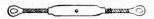 

FIG. 30.—Bottle Screw.   FIG. 31.—Bottle Screw Preventer.

Bottle Screw Preventers.—Constant vibration may sometimes cause bottle screws to "walk back" or slacken up, and to stop this either preventers are fitted or the screw is seized with seizing wire.

Care of Bottle-Screws.—To ensure that bottle screws will be in working order when required it is essential to protect all exposed threads. In the case of standing rigging the practice is to smear plenty of tallow over the threads, then serve over with rope-yarns until level with the edge of the bottle. Canvas gaiters or covers are sewn over the yarns and these, when painted, give adequate protection from the weather.

It is essential, too, that both holes in the bottle should be plugged to prevent water percolating inside.

Sometimes the ends of both screws come together directly opposite the hole in the bottle, and these will quickly rust if not protected, giving endless trouble when they are ready for coming adrift.

Brow.—Heavy gangway used as a means of access to ships in dock.

Bull Rope.—Length of rope (usually 3-in. or 4-in. stuff) used for dragging cargo out to the sides or ends of a hold.

It is also the name generally used for the light wire attached to topping-lift spans, and with which the derrick is hove up. With this type of gear the topping-lift span is shackled to a short length of chain which is also shackled to the deck, and the bull rope is only used for heaving up or lowering the derrick.

Bunting.—The material of which flags are made. It was composed wholly of coarse wool, and manufactured in 18-inch widths but is now a wool/nylon mixture.

Burlap.—Coarse bagging material used for covering ironwork before loading grain, and for many other purposes.

Cane Fenders.—When a vessel is laid against a dock wall for any length of time, cane fenders are hung over the side to protect the hull. They are made from bundles of light canes bound tightly together with wire.

Canvas Bucket.—Used almost exclusively when taking sea-water temperatures. It has a wooden bottom, and is sometimes weighted on its top edge to make it tilt more easily when dropped over the side.

Cement.—Two of the principal uses to which cement is put on board is the sealing up of spurling pipes, and the making of cement boxes to stop leaky rivets.

Chafing Mat.—A mooring rope or wire led across the stem is likely to become damaged through chafe. To prevent this a square of sword matting is first hung in position, the rope or wire is hove tight, and the mat then takes the chafe.

Chain Gantline.—Although sometimes used with funnel boards, chain gantlines are not popular for lowering purposes. They are used principally as Dummy Gantlines.

Chain Stopper.—A short length of chain (about 6 feet) with a rope tail spliced in one end and a large link or shackle fitted in the other. It may be either shackled to a ringbolt or rove around one of the bollards when "stoppering off" a wire.

Chutes.—Small steamers always have a chute amidships, so that refuse thrown overboard will shoot clear of the ship's side. When alongside a wharf, big wooden chutes are hung over any scuppers or soil-pipes which are likely to spout water over the quay. They are heavily weighted on the bottom to prevent the wind swinging them about. Weighted canvas screens called "scupper cloths" are often used for the same purpose.

Cork Fenders.—When a vessel appears to be in danger of grazing a dock wall, a cork fender is lowered down to the point of contact in order to lessen the impact. The outside is made of coir rope laced tightly over the round canvas bag containing broken cork.

Cowl Covers.—A ventilator switched back to wind will often draw spray down the cowl and, in cases where a ventilator cannot be swung off the wind, the opening is protected by a canvas cowl cover.

Dodgers.—Canvas screens which gave protection from the wind in exposed positions, principally the navigating bridge, forecastle head and crow's nest. The name now applies to permanent fittings of wood or steel.

Dressing Lines.—Lengths of small-size wire to which flags are seized with sail twine when the intention is to dress ship. They are made and kept specially for the purpose, and are fitted with swivels and downhauls.

Dummy Gantline.—When a masthead sheave or funnel block is difficult to reach, an old piece of rope, wire or chain, is always left rove in place.

When work has to be done aloft, it is then only necessary to "marry" the new gantline to the dummy and pull it through the sheave.

Eyelets.—There is no standard method of indicating the various sizes of brass eyelets used in canvas work, so some people simply order "small", "medium" or "large", in which case the ship-chandler uses his own judgment. The most popular method adopted by many firms is the use of numbers indicating a certain sized hole.

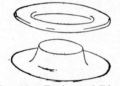

Fig. 32.—Eyelet and Ring.

As the numbers vary among different manufacturers, the safest plan is to order by a measurement which indicates the diameter of the hole required, when the eyelet is closed. Fig. 32.

Eyeplate (*See* page 206) "Monkey Face".

Firebars.—Heavy iron bars fitted to the inside of ship's furnaces. Broken ones are occasionally used on deck for any purpose where

Fig. 33.—Firebar.

weight is required, such as ballast for boats, weights for scupper cloths, etc. Fig. 33.

Firebricks.—Bricks of the ordinary size and shape, but made with special fire-resisting qualities. They belong to the engine-room department, but are sometimes handled by the sailors.

Flag Clips (or **Englefield Clips**).—Small elliptical metal rings which connect by means of V-shaped openings in their sides. They enable flags to be quickly bent together, or to halliards, and without any danger of them coming adrift afterwards. There is also a swivel on each clip which prevents turns getting into the flag halliards.

Flag Toggles.—Small round pieces of wood about $1\frac{1}{2}$ inch long. These are spliced to a flag for connecting it to others when clips are not fitted.

Forelock.—A flat, wide locking-piece which fits into a slot in the pin of a shackle. This prevents the pin of the shackle from withdrawing. In order that the forelock will not fall out the "legs" are opened sufficiently to jam them in the slot of the pin.

Funnel Boards.—Four light planks, cut to such a size that they can be bolted together around the funnel to act as a staging. They are hoisted to the top by means of four gantlines—one at each corner.

Gantlines.—Masthead gantlines are invariably cut from a coil of $2\frac{1}{2}$ inches, and should be long enough to reeve through the masthead sheave and reach the deck again. Lowermast gantlines are long enough to reeve through a tail block or lizard hung in the eyes of the rigging. They must be kept dry and in good condition as they have to bear a man's weight when painting the masts. Funnel gantlines are always unrove as soon as possible to prevent injury through heat.

Guest (or **Guess**) **Warp.**—A rope of suitable length to reach from well forward to foot of the accommodation ladder, so that boats coming alongside will have something to catch hold of and make fast to.

Hatch Battens.—Long, flat iron bars which hold the tarpaulin hatch covers in place. They fit into cleats and are wedged tightly in position with wooden wedges.

Hatch Covers.—Heavy wooden covers usually referred to as "Hatches". They are numbered and fit over the beams of the hatchway. These in turn are covered with tarpaulins to make the hold watertight. Now that the New Load-line Regulations are in force, the wooden hatch covers are rapidly being superseded by the MacGregor-type watertight steel covers which allow a vessel to carry more cargo by being allowed to load deeper.

Hatch Tent (or **Hood**).—A large tent-shaped hood which is hoisted up on the cargo runner of a spare derrick and made fast at the bottom to the coamings.

This permits cargo to be "worked" during rain. When loading chilled meat a much larger tent is used. It completely covers the hatch and derricks as far as the ship's side, and ensures that the meat remains chilled while loading.

Heaving Line.—Coil of "small stuff" used for throwing on to a dock or quay wall, in order to establish contact when a vessel is berthing. By means of these lines the mooring ropes are hauled ashore. It usually consists of a coil of about 15 to 20 fathoms of 15- or 18-thread ratline which is weighted at the end (to carry the line well when thrown) by means of a "heaving line knot" or a "monkey's fist" (*see* Knots.)

Jumping Ladder.—Light, handy ladders which are easily carried about the decks to wherever men are painting overside. But for these it would be necessary to climb the stage ropes to come on board.

Ladders.—A number of small, wooden ladders are generally carried for reaching the tops of davits, ventilators, etc., when painting. In addition, it is usual to carry a long one of sufficient length to reach down into the holds or to the wharf. They are usually kept in the hold, or on the boatdeck.

To Hook Tackle on Lanyard.—Stretch the lanyard tight and hold in left hand. Lay back of hook on lanyard and take two round turns on hook itself. Twist hook so that the standing part of the lanyard will come inside the point and lay over the two turns. When the strain comes on, the lanyard should jam the turns.

Lizard.—A lizard is a short length of rope or wire with a thimble spliced in one end. This thimble is used instead of a tail block or sometimes instead of a lead block.

Long Arm.—See Manhelper.

To Fix Brush on Manhelper.—If a manhelper is to be of any use, the brush must be put on very tightly or it will give endless trouble.

First, lay the brush in the slot at the proper angle and put on a good cross seizing. Next, make a yarn fast at the top of the handle and pass it across to the manhelper at a wide angle. Take two such turns, and frap the parts down to the manhelpers before making fast. Instead of frapping, a small piece of wood is sometimes used as a Spanish windlass, and this is made fast in place when tightened up.

FIG. 34.—Manhelper.

Marine Glue.—A black substance resembling pitch. It is used as a filling between deck planks after caulking, and is derived from a bitumen and also contains glue. Steam yachts use a white variety for the same purpose.

Megaphone.—A cone-shaped speaking trumpet which magnifies the voice considerably. It was used when hailing anyone a long way off, but is now superseded by a public address system.

Messenger.—When a mooring rope or wire is too big to go around a winch barrel, a smaller line (about 3 inches) is made fast to it, and the big one is hove in bit by bit. The smaller one is known as a "Messenger".

"Monkey's Face".—Thick, triangular-shaped iron with a hole in each corner. It unites the Topping Lift Span, Bull Rope and Chain Preventer. Is sometimes called an " Eyeplate".

Mooring Ropes.—The size and number of mooring ropes will depend upon the size of the vessel. As a general rule not more than four will be found at each end. The size will be in the neighbourhood of 6, 7, 8 or 9 inches circumference or 48 to 72mm diameter.

Mooring Wires.—As with mooring ropes, no hard and fast rules can be given for wires. The average steamer has two or three at each end and they are used as headwires or backsprings. In addition, there is usually a very heavy one known as an "insurance wire", but this is only used in an emergency.

Oakum.—Old rope-yarns teased out and tarred. It is used for caulking purposes and is hammered tightly into the deck seams. Hot pitch or marine glue is poured in to make the seam watertight.

Paint Drums.—Sheet metal containers used for the carriage of paint. They are made in a great variety of sizes.

Painting Flat.—To avoid using lifeboats when painting overside, an oblong punt about 8 feet long is sometimes carried for that purpose. They are usually open on account of weight, but those used in docks by shore workers are invariably decked over.

Paint Strainer.—An article resembling a saucepan or basin with a wire gauze bottom. It is used for straining off skins or other foreign matter from paint.

Pilot Ladder.—A rope-ladder with wooden steps. It is hung over the ship's side for pilots and others who wish to board a vessel at sea. *See* page 86.

Rat-guards.—Large circular metal discs which completely encircle a mooring rope or wire. Their purpose is to prevent rats

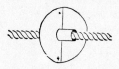

FIG. 35.—Rat-guard.

coming on board or going ashore. They measure about 2 feet 6 inches in diameter, and must fit snugly and have all holes blocked if they are to be effective. Fig. 35.

Ridge Wires.—Wires rove through awning stanchions at the ship's side and set up tight with bottle screws. Awnings are made fast with stops to the ridge wires.

Rope Stopper.—About 2 fathoms of 3-inch or 3½-inch manila with an eye spliced in one end. It holds the weight of a mooring line when transferring it from the winch barrel to the bollards. Small ones are used in the same way with tackles and falls. *See* Chain Stopper.

Scupper Cloths.—*See* Chutes.

Selvagee Strop.—An ordinary strop will not always grip when a strain is applied to it, in which case a selvagee strop is made with a bunch of rope-yarns. The rough surface of the yarns is more liable to grip when a strain is put on the strop.

Senhouse Slip.—Simple releasing arrangement used on all boat gripes. Fig. 36.

FIG. 36.—Senhouse Slip.

Separation Cloth.—Coarse bagging material or burlap used for separating different parcels of the same cargo.

Shores.—Heavy pieces of timber. They are used to support an erection such as a bulkhead which is showing signs of collapse, or to make a new erection more secure.

A "Tom" is the same sort of thing on a much smaller scale.

Split Pin.—This may be described as a safety pin. Its function is to prevent a nut unscrewing, or a bolt withdrawing from its proper place.

Stage Clamps.—Another name for Stage Hangers. Fig. 37.

Stage Hangers.—An iron gadget shaped to hang over the teak rail of the bridge, and from which a stage is hung when work has to be done on the fore part of saloon houses, etc.

They are usually parcelled over to prevent scratches appearing on the varnished surface of the rail.

Stages.—Light planks suspended over the ship's side whereon a sailor stands while painting. It is fitted with a horn (crosspiece) at each end to keep it well off the side and allow room for movement.

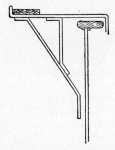

Fig. 37.—Stage Clamp.

Stem Fenders (or Goole Fenders).—These are similar in shape and size to a lifebuoy. They are compulsory when entering or leaving the port of Goole, and are hung across the stem, near the water's edge on account of the low quay walls in that port. Regular traders are also provided with a number of long fenders which reach from the deck to the water's edge. They are made in the same way as round sennet.

Striker.—*See* Manhelper.

Spanish Windlass.—This is a very handy arrangement whereby added power is gained when heaving parts of a rope or anything else together.

It is a simple system employing leverage, and can be used on such jobs as making cane fenders, where the cane has to be bunched tightly together, or when stropping blocks, and the strop has to be squeezed in tight before the seizing is applied. Various methods may be adopted when making a Spanish windlass, but the following way is as good as any:

Say two parts of a rope are to be hove together. First, lay a crowbar across the two ropes where the heaving is to be done. Next, with a good strand of yarns take a round turn (two may be necessary) around both ropes, so that the ends come up alongside the crowbar. An eye is then formed on the end of each strand, either by twisting round its own part or with a docker's splice. A spike is then placed through each eye, and both are twisted around the crowbar until the two parts of rope come together. If the strand is well greased beforehand, the heaving operation will be much easier.

Spurling Pipe Covers.—Canvas covers which are meant to prevent water finding its way down the spurling pipes and into the chain lockers. Actually, they are seldom used; instead, the pipes are plugged with old bags and covered with a layer of cement. This is easily broken out again when the anchor is to be "let go".

Sword Matting.—Trade name for machine-made matting, narrow widths of which are used for covering guard rails of ladders.

Wide strips are also used for griping boats to a griping spar, while swung out at sea. The wide, flat surface is not so liable to chafe as with other materials.

Sword matting is seldom kept in stock but is made to order according to the size required. The only real demand is for sizes between 4 and 8 inches. What is known as ordinary quality is made from hemp, and may be obtained either tarred or untarred.

Tackline.—About 6 feet of halliard with a swivel clip at each end, used for separating groups of flags hoisted on the same halliards.

Taking Temperatures.—On many ships it is the practice to take temperatures of the sea-water every four hours. In some cases this is done to determine the proximity of such currents as the Gulf Stream, while in other vessels, such as fruit or frozen-meat carriers, it is a helpful guide when the holds have to be kept at a certain temperature. A canvas bucket with a long lanyard is dropped over the side well clear of any engine-room discharge, and a thermometer is left for a few minutes in the water thus obtained.

(See Appendix, page 304.)

Toggle.—A piece of wood which provides a quick method of making fast or releasing a rope. The size of a toggle depends upon the purpose for which it is intended. For instance, if the eye of a mooring rope were passed through a mooring ring, it would take a substantial piece of wood, shoved through the eye, to form a toggle.

On the other hand toggles as small as one inch in length will be found on the hoist of a flag for attaching it to the flag halliards. A toggle about 8 inches long is always used when connecting a lazy painter to a lifeboat, and by this means the painter can be slipped very quickly.

Tom.—Small shores used to wedge anything into position.

Tow-rope.—Many ships have a special tow-rope which is kept specially for that purpose. The usual practice is to have a good length of wire spliced to a short end of coir, manila or nylon. The rope acts as a "spring" and is often of the cable-laid type.

Trimming Shovels.—As it is very difficult to dig down into a heap of coal with a square shovel, the pointed variety is used for this purpose. They are known both as trimming shovels and round-nosed shovels.

Water Service Valves.—Carelessness when operating a water service valve may easily result in a damaged pump, so the arrangement in each particular ship should be understood before touching it. The principal thing to remember is that before water is turned off from a deck hose, an outlet must be provided for it elsewhere. This is done by opening another hydrant on the service pipe.

Switch Valve.—The switch valve is usually situated amidships at the place where the pipe, which conveys water from the pump, joins the service pipe.

Two-way Valve.—Water can only be switched to flow forward or aft. No outlet overboard is provided. If the hose is in use on the foredeck, and is to be shifted to another connection, a "cap" must first be removed on the afterdeck to allow the water to escape. The water is then switched over in that direction until the hose is disconnected forward.

Three-way Valve.—With this type water can be switched outboard while shifting the hose.

Ventilator Plugs and Covers.—Ventilator cowls are nearly always unshipped and stowed away during heavy weather, the exposed opening being sealed with a wooden plug, over which a canvas cover is lashed. In the event of fire the ventilators are plugged to help in smothering the fire.

Whiting.—Chalk ground and purified. It forms a good putty when mixed with about 20 per cent linseed oil.

Winch Covers.—These are seldom seen except on passenger vessels, and only when winches are placed in such a way that passengers might come in contact with them.

Wind Chute.—Sheet-metal chute which fits into a porthole and scoops air inside the compartment.

Windsails.—Long, canvas ventilators with a top section adapted to catch wind and shoot it down the engine-room, or any other place over which it is hoisted. Oil-tank steamers have one to each tank to blow out gas fumes. They are always fitted with wooden hoops to keep them cylindrical in shape. Largely superseded by forced draft systems.

Wire Brooms.—Small brooms with wire bristles. They are used for scouring loose rust from ironwork, and marine growth from the ship's side, particularly near the waterline.

Wire Grips (or Bulldog Grips).—Instead of splicing an eye in a wire, the end may simply be doubled back and held in place tightly with three or four wire grips. Fig. 38.

Fig. 38. Wire Rope Grip.

CHAPTER VIII.

PAINTING

Preparation of Surfaces. Mixing Pigments and Binders. Brushes. Painting Derricks, Funnels, Topsides and Deck Fittings. Varnishing and Enamelling.

Painting at Sea.—Much of the painting done at sea is simply slap-dash work. There is never time to put paint on properly, and still less time to wait for weather conditions. The result is that blistering, peeling and bleeding is a quite frequent penalty for hurried work.

The surfaces, too, are very often dirty and greasy, as well as rough, through successive coats of unstrained paint, not to mention accumulations of rust, dust, etc. In some parts of the ship, such as over the side, this cannot be avoided or rectified, but there is no reason why a little more time could not be spent on deck houses, alleyways, and so on.

Such places catch the light and show themselves to advantage when finished off properly.

The principal reasons for putting paint on any surface are preservation and decoration, and the only way to attain these objects is to make good use of a scraper, pumice-stone, red- or white-lead, and apply the final coat on a smooth, clean, dry surface.

A sailor is not expected to be a first-class painter and decorator, but he requires to have more than a nodding acquaintance with paint, enamel and varnish, if he hopes to apply them successfully.

A knowledge of the materials and how to mix them will enable any amateur to turn out better work than the man who is only familiar with brush work.

Composition of Paint.—All paint consists of four essentials. These are: the pigments, the binder, the thinner and the drying agent.

Pigments.—The principal pigments are white-lead, zinc, oxide and red-lead. White-lead is used more than any other type of pigment in the composition of ordinary paint; it gives a good body, whereas zinc oxide is poor in this respect. To offset this lack of body, zinc is considered to be more durable, and it retains its pure colour better than lead. White-lead improves with age and provides a tough, elastic film over any surface. It expands and contracts with changes of temperature, and remains uncracked to the end.

White-lead and zinc pigments are supplied in oils, and they only require to be reduced with oil and turpentine. Red-lead is supplied in powder form. It is used extensively as a first coating on ironwork, dries quickly, and forms a good, hard coat.

Binder.—Oil is the substance used to bind the pigment down to the surface. A variety of oils—mostly vegetable—may be used for this purpose, but for general suitability, reasonable price and good drying qualities, boiled linseed-oil is the best.

Gold Size.—A "binder" used a good deal on fine work. It is made from very pure gelatine.

Thinner.—The "thinner" is added to paint to promote ease of application by bringing the composition to a more fluid state. Genuine American turpentine is the best paint thinner, but a substitute such as white spirit—a derivative of petroleum—is invariably used on board ship. All thinners must be water white and free from solid matter.

Driers.—A quantity of driers is usually added to paint to accelerate the drying process. It speeds up the chemical reaction on which drying depends, and lessens the time required for the thorough hardening of a painted surface. It acts on the oil, but not on the pigment, and has no protective or decorative value. Many types of driers are in use, some of a fluid nature, and others the consistency of putty. Most of them, however, contain compounds of lead and various types of manganese. An excess of driers is definitely deterimental to the life of paint.

Terebene.—A much-used drier made from raw linseed oil and gum copal. It also contains Litharge, which is a monoxide of lead.

Red-lead.—This is sometimes used as a drying agent when oiling wooden decks. Mixed as paint it dries very quickly.

Patent Composition Paints.—Anti-corrosive, anti-fouling and boot-topping paints are more expensive than ordinary prepared paints, and they are manufactured specially for the underwater parts of a ship's hull. They have a heavy pigment which requires to be constantly stirred to prevent settling while in use. In fact, with some makes it is necessary to turn the paint drums upside down periodically to prevent settling while in storage. Evaporation is so fast with these compositions that the paint is nearly dry as soon as put on, with the result that a vessel may leave a dry-dock without unnecessary delay.

Anti-fouling paint contains a large quantity of oxide of mercury which is injurious to bare steel and will set up corrosion; therefore, an anti-corrosive is put on first. It acts as an insulator against the anti-fouling which is painted over it. This is put on to retard the growth of barnacles, grass, etc.

Boot-Topping.—The area between the light load-line and load-line proper is known as the boot-topping. As it is very often out of water, it can be scrubbed when necessary to remove all marine growth; therefore, an anti-fouling (and anti-corrosive) is unnecessary.

Instead, a composition known as boot-topping is applied to that section. This is a less expensive paint which, like the other two, has a heavy pigment and quick-drying qualities.

Note.—Most composition paints can be thinned down with naphtha.

Bituminous Paints.—These are considered a good preventive against corrosion and are used extensively on steelwork, especially under water. They are composed of tar mixed with resin and tallow.

Funnel Paints.—The essential difference in these paints is that they must be impervious to heat, or blistering and cracking will take place, leaving unsightly patches on the funnel.

MIXING PAINT.

Most of the paint sent on board ship is all ready mixed and only requires a little thinning before use. This applies particularly to funnel and mast paints, whereas, that used for deck houses and the ship's side is usually mixed on board. The consistency of such paint depends largely on the amount of pigment and oils available, which in turn depends upon the class of vessel and the trade she is engaged in, with the result that no hard-and-fast rules can be given in connection with paint mixing, except for a few broad principles which are again subject to individual taste.

A bo'sun or lamptrimmer very seldom mixes paint by weight. As a rule they draw on their former experiences to estimate the quality and quantity which will be required.

Much depends on the class of pigment and the surface to be covered. Cheap pigments cannot be mixed as thin as the better-class ones. Similarly, dark paints can be mixed much thinner than any other colour and still retain the same concealing power when applied.

Then again, one must bear in mind the effect of the climate. For instance, paint in hot climates does not require so much drier as in a moist one, and paint intended for a fish plate would require more weather-resisting qualities than a stokehold bulkhead below decks.

Covering Power.—Only a rough estimate can be made of the area a given amount of paint is likely to cover, particularly where the ship's side is concerned. A rough surface requires a good deal more paint than a smooth one, but as a general rule, somewhere about 6 or 7 lbs. of paint will cover roughly 50 sq. yards. Black paint can be thinned down a good deal, and will cover a larger area than any other given quantity of paint.

Mixing White Paint.—Boiled oil must never be used for white paint or it will dry a creamy colour, but it is quite safe to use it for all darker shades of paint.

For white paint, raw linseed oil and turpentine should be used in the proportion of one quart of turpentine to one gallon of raw linseed oil.

Both white-zinc and white-lead or a quantity of both may be used in the making of white paint, but of the two the zinc gives the whiter colour. The following is an average mixture: 7 lbs. white-lead or white-zinc (or both); $\frac{1}{2}$ gall. raw oil, $\frac{1}{4}$ pint turps and 2 ozs. driers.

This mixture may be varied a good deal, and is only given as a rough guide. For larger quantities it is only necessary to double or treble the given figures as many times as desired.

It is a common practice to mix up a large quantity of pigment and oil, then add the turpentine to each pot of paint just before it is taken away to be used. By doing this it is possible to vary the consistency of the paint to suit different jobs.

White paint improves with standing, and it is usual to mix it up some two or three weeks before it is to be used, giving it a stir up now and again.

It is better not to add the driers until ready to use the paint, otherwise skins will form very quickly, and every potful will require to be strained.

Another common practice is to stir a small quantity of blue (blue pigment mixed with raw oil) into white paint. This has a "bleaching" effect, and neutralises any tendency of the paint to turn a creamy colour on account of the oil content.

Mixing Red-lead.—This pigment has such good drying qualities that driers are not usually added. It forms a good, hard coat, and is used extensively to cover bare iron, or any places where rust is likely to show through.

The following will make up one gallon of red-lead ready for use. 20 lbs. red-lead, 6 pints linseed oil, 2 gills turpentine.

To prevent excessive settling of the lead in the pot when in use damp the powder with boiled oil, and let it stand for half-an-hour before mixing.

Mixing Flatting.—This may be prepared from white-zinc, turpentine, and a small quantity of gold size to act as a binder.

Mixing Mast Colour.—$1\frac{1}{2}$ cwt. mast colour, 3 gal. boiled oil, $\frac{1}{2}$ gal. raw oil, $\frac{1}{2}$ gal. terebene, $\frac{1}{4}$ gal. turpentine, $\frac{1}{4}$ gal. varnish.

Mixing Red Funnel Paint.—Take the required quantity of red-lead, dry it in the sun for half-an-hour or so, then damp with paraffin-oil, so that it is even drier than a paste. After standing for half-an-hour, thin down to the required consistency with colza oil. This gives a light red.

Mixing Black Funnel Paint.—To the stiff black paste add 1 part turpentine, 3 parts boiled oil, and patent driers about the size of an egg. For a deep black, add drop black broken up in turpentine. Varnish must never be added to funnel paint containing spirit.

Mixing Colours.—

French Grey	White-lead and prussian blue tinged with vermilion.
Regulation Grey	White-lead and black in proportion of 16 to 1 (7 lbs. black—1 cwt. white-lead).
Silver Grey	Lampblack and indigo.
Pearl Grey	Black, blue and white.
Pearl	Blue and lead colour.
Lead Colour	Lampblack and white.
Straw Colour	Yellow and white.
Cream Colour	Chrome yellow, best venetian red and white-lead.
Flesh Colour	Crimson lake, white and vermilion.
Salmon Colour	White-lead tinged with venetian red.
Pink	Carmine and white.
Rose Colour	White and crimson lake.
Fawn	Burnt sienna and white-lead.
Brown	Black and red.
Chestnut	Brown and white.
Chocolate	Venetian red and black.
Buff	White, yellow and venetian red.
Drab	Raw or burnt umber and white-lead with a little venetian red.
Dark Green	Light green and black.
Pea Green	Green and white.
Brilliant Green	Emerald green and white
Olive	Red, blue and black.
Orange	Yellow and red.
Purple	Blue, white and crimson lake or vermilion.
Violet	White-lead, ultramarine, vermilion and a little black.
Imitation Gold	White-lead, chrome yellow and burnt sienna.
Stone Colour	White-lead, raw sienna and umber.

Drying of Paint.—Paint dries and hardens through absorbing oxygen in the air, a drying effect produced by linseed oil, and the evaporation of turpentine. The time spent in oxidation may be speeded up by adding special driers to paint, but the use of heat for the purpose is useless.

Blistering.—Unsightly blisters will invariably arise on woodwork painted during damp weather, or on wood not properly seasoned and full of sap. A certain amount of moisture in patent driers, or dampness imprisoned between coats of paint, may be the cause of blisters forming on ironwork which has been painted. Dark-coloured paints blister more easily than others as they absorb more light and heat.

Tacky.—If a finger placed lightly against a recently-painted surface adheres slightly to it, it is said to be "tacky". This con-

H

dition may last for some time if the paint has been put on over a dirty, greasy surface, or if there is an excess of driers in the paint. In the case of varnish, insufficient ventilation would have the same effect.

BRUSHES.

Animal bristles, vegetable fibre and horse-hair are some of the materials used in the manufacture of paint-brushes, while varnish brushes contain bear's fur, and those used for graining are made with badger's hair.

Bristle brushes are the best and are easily identified by their natural spring, and the fact that the bristle ends are tapered.

Also, if a strand is pulled out and bent double, the fibre strand will usually crack, while the animal bristle will go back into its normal shape. Or again, a match applied to a strand would cause fibre to burn, but bristle would not.

New brushes, or those which have not been used for some time, should be soaked in water, otherwise the bristles will come out very freely, but this does not apply to rubber-set brushes.

Those used for paint and varnish are mostly set in either cement or rubber, and are usually bound with copper wire, or cased in metal ferrules of copper, aluminium, or tin.

The bristles of those made for use in varnish are always bevelled, and are set in rubber to render them impervious to the action of spirits in the varnish, turpentine, petrol, etc.

Those set in cement are meant to be used in paint, and are never guaranteed to resist the action of spirits, which will dissolve the cement and ruin the brush.

Breaking-in Brushes.—New brushes of the cheaper variety seldom "stroke off" neatly, therefore, to break them in they are used on a rough surface for a few days. The ship's side is best for this purpose. When the ends of the bristles are nicely tapered and the brush in good condition, it is then washed out and kept ready for any work which requires a good finish.

Binding or Bridling Brushes.—New round brushes must be bridled or bound to reduce the length of the bristles, and although some are provided with a patent metal bridle, the usual practice is to bind it either with rope-yarns or bunting. When bunting is to be used, take a piece long enough to reach from the foot of the handle to the end of the bristles, then roll it round the brush twice. Fig. (1a). Pass a seizing about half-way down the bristles, turn the bunting back and make fast with another seizing around the foot of the handle. Fig. (b). Never bridle a brush too tightly or the bristles will lose their spring and become distorted.

Working In.—Before starting work all surplus water must be shaken out of the brush. Next, dip it well into the paint and rub it around the sides of the pot, or on a spare piece of wood, in order that paint will find its way right through the bristles. This is important after a brush has been washed out in case any traces of the old colour remains.

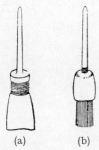

(a)   (b)

Fig. 1.—Binding Brushes.

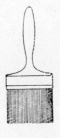

Fig. 2.—Flat Brushes.

Flat Brushes.—Most seafarers consider them to be the best type of brush for all general work. Sizes are according to width in inches, and they range through each half inch from $\frac{1}{2}$ inch to 4 inches. Fig. 2.

Fig. 3.
Round Brush.

Fig. 4.
Dog's Leg Brush.

Fig. 5.
Sash Tool Brushes.

Round Brushes.—Most of the brushes which come under this heading are really oval in shape, and the sizes are determined by numbers. They run from $\frac{1}{0}$ to $\frac{8}{0}$. Fig. 3.

Dog's Leg.—A very handy and popular type of long-handled paint-brush.

Sash Tools.—Small, round brushes used for reaching corners where a larger brush cannot be worked, (the name comes from sash windows ashore). They are usually string-bound, but are also found encased in ferrules of aluminium, copper or tin. As regards size, seafarers usually class them as "small" and "large", but each has a number, the complete range of which is from No. 0 to No. 14. Fig. 5.

Fitches.—These are always used for such jobs as painting in very narrow holes and corners, and also for painting names, numbers and other small work. Although usually designated "small" and "large" there is a big range of numbers extending from No. 1 to No. 16. Fig. 6.

FIG. 6.—Fitch.   FIG. 7.—Lining Tool.   FIG. 8.—Pencil.

Lining Tools.—Brushes specially designed and cut for drawing lines, but only used on board for painting figures or names. Sizes run through each 6mm from 6mm to 36mm. Fig. 7.

Pencils.—Small, sharp-pointed brushes meant for artistic work, but employed in painting numbers and letters. Fig. 8.

Tar Brushes.—Two types in three sizes (No. 1 to 3) are in general use. One has a short handle similar to a paint-brush, and the other type has a long handle about 4 feet long. Fig. 9.

Whitewash Brushes.—These are invariably "two knot" brushes. That is, the bristles are bound into two knots, and this gives two brushes on the one handle. Sizes are by weight, and these range through each ounce from 8 ounces to 12 ounces. Fig. 10.

Stencil Brushes.—Although seldom found on board the average steamer, these brushes are worth the small extra cost. They are about the same size as a shaving-brush, but the ends of the bristles are cut off to give a flat surface. Sizes are in inches and indicate diameter of bristle surface. The range extends from about 12mm to 36mm. Fig. 11.

Care of Brushes.—The method in which brushes are stowed away when not in use will determine their future usefulness. The usual practice is to stand them in a tin of water, but this will cause them to become crooked and twisted—especially those of an inferior quality. It is better to suspend them in the water, not only to preserve their shape, but to keep the bristles free from grit in the bottom of the tin.

Brushes set in metal should not be immersed in water, but placed in linseed oil, or a mixture of oil and turpentine.

Washing Brushes.—If not to be used for some time, it is better to wash a brush out with turpentine, soap and water, but on no account use soda or this will ruin the bristles.

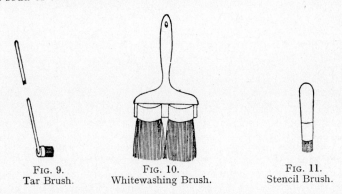

Fig. 9.   Fig. 10.   Fig. 11.
Tar Brush.   Whitewashing Brush.   Stencil Brush.

Whitewash Brushes.—Wash out in cold water after use, and after straightening bristles, hang up to dry with a string rove through a hole in the handle. Before using, both new and old brushes should be soaked in cold water for one hour to tighten the bristles in case they have become loose through shrinkage in dry weather.

Varnish Brushes.—As dampness is the great enemy of varnish, the brushes should never be kept in water, so for this reason they are always kept in oil, varnish or turpentine, or a mixture of all three.

Softening Brushes.—If, through neglect, a brush has become hard and stiff with paint, it can be softened again by immersion in a paint remover such as Pintoff. This will take some two or three hours and should be followed by washing in turpentine, then in soap and water.

Paint Rollers.—Paint rollers are often used instead of brushes, particularly when painting large, flat surfaces. Rollers may be lambskin or nylon covered and should be used in conjunction with properly designed paint trays into which the paint is poured. The tray is on the tilt and after dipping the roller into the paint, it is run up and down on the dry part to ensure an even distribution across the whole of the roller besides removing excess paint.

Rollers are supplied with short or long handles and provide a very quick method of covering large surfaces but if used without a tray can also be a somewhat wasteful method of painting.

After use the roller should be cleaned with the same thinner as used for thinning the paint, washed and rolled out as dry as possible, being fully dried before storing. If it is in day by day use it can be treated the same as a brush and left to soak.

Spray Gun.—In this case the general principle is that air-pressure is used to spray a paint-saturated mist through the nozzle of the spray gun by a trigger control. For good results it is essential to keep the same distance between the nozzle and the work to be painted and to keep the spray gun in the vertical position, i.e. the gun must be moved up or down or from side to side parallel to the surface being painted. It is also most important that the surface is clean and dry to ensure proper adhesion. It should only be used in a still atmosphere and after use must be thoroughly cleaned. The health of the operator must be safeguarded bv providing a mask and goggles together with overall clothing.

PREPARATION.

Any surface which is to be painted must be clean. Coal dust and such like are liable to work through paint—especially white—and cause discolouration. Grease in particular should be cleaned off with paraffin or unsightly streaks will result. Rust marks must also be removed or they are liable to show through. If sugi-mugi fails to erase them, a rub with bath-brick will probably prove effective. To obtain that really finished look, a surface must be smoothed off with pumice-stone, and, if necessary, the paint should be strained, although this is seldom done except with paint intended for bridge or saloon houses. When loose rust is encountered it must be scraped away or it will flake off on its own later and leave unsightly bare patches. Blisters must also be cut away and the edges fined down with a knife.

Touching Up.—All bare places which have been scraped or cleaned off—especially where bare iron shows through—must be touched up with either red- or white-lead. A thick coat should be applied, and if red-lead is used, it will require a coat of paint to tone the colour down before the final coat is applied.

Bare Steel.—Bare steel plates should be smoothed off with a scraper and given a coat of boiled oil. One or two coats of red-lead is usually followed by two coats of the final colour to finish off. That is: one for toning down the red-lead, and one more when adjacent plates are being done. To determine whether a plate requires one or two coats of lead, much would depend on whether the place is in an exposed position or whether it stands a lot of wear and tear.

Bleeding.—When rust breaks through paintwork, and ugly yellow streaks and blotches appear, the affected spot is said to be "bleeding". The usual remedy is to scrape away the paint and accumulated rust, apply either one or two coats of red- or white-lead and finish off with a coat of the surrounding colour. In addition, if available, a thick coat of varnish next to the steel is an even better cure.

Preparing New Wood.—All new wood must be given a priming coat. This fills up the grain of the wood and gives a colour foundation for the final coats. Priming coats are usually made a bit thinner than other paints. On new wood it is also advisable to use Patent Knotting which prevents the knots showing through. Only the knots themselves are touched up, and being a quick-drying material (about 10 minutes), two coats are usual.

Putty.—This is supplied ready made to many vessels, particularly those where it is used for making gun port doors watertight. When not supplied it can be made as under, but as whiting is seldom to be found on the average ship, the white- or red-lead variety is mostly used.

Ordinary Putty.—Is made from whiting mixed with about 20 per cent. raw oil, and is used principally for filling cracks or holes in woodwork. It may be tinted with practically any colour to suit whatever paint is to be applied over it.

White-lead Putty.—Ordinary white-lead stiffened up with whiting. It is often used without the stiffening, especially where the intention is to stop rust from forming.

Red-lead Putty.—White-lead mixed with red-lead in powder form. If necessary, it can be slackened with a little raw oil. This mixture is a good one to use on gun port doors where the action of the water will wash away the ordinary types of putty.

Filling.—Another good filling, which can sometimes be used in place of putty, is a mixture of cement and raw oil. It can be smoothed off easily to take a coat of paint, and is impervious to heat, with the added advantage of setting much harder than putty. It can also be used as a substitute for marine glue.

Keane's Cement.—A good filling for levelling off uneven surfaces, particularly on deck houses or bulkheads where a good surface is required. Its chief virtue is that it hardens very quickly and does not crack and break away. Mix the powder into a paste with three parts turps and one part gold size. Wash surface or part to be filled with turps and apply paste. When smoothing off sprinkle turps on the surface if necessary.

APPLYING PAINT.

Paint should be applied as thinly and evenly as possible. To do this properly it must be "crossed" before "stroking off". That is to say, it should be spread across the direction which the finishing-off hairstrokes will take.

Needless to say, "holidays" (parts left dry) should be avoided or they will give an unpleasant effect when touched up afterwards.

It is a good plan to coat over all rivets, bolts or other obstructions before filling in the flat parts. One reason for this is that, with quick-drying paints, a large area can be filled in and stroked off before it sets, otherwise a patchy appearance will result where the dry and wet sections meet. A sailor should develop the habit of planning out in his mind the painting he is about to do.

Hairstrokes.—The best guide to the direction which hairstrokes will take is the old paintwork on which they are always visible.

If in doubt, it should be remembered that the ship's side, deck, and skirtings at the bottom of saloon houses are the only places where they are likely to run fore-and-aft. Elsewhere, they are either up and down or athwartships. Some vessels may differ from this, but this is the usual practice.

While on the subject of hairstrokes, it is as well to note that the proper way to paint a panelled door is not to simply start at the top and work down. Do the beadings, panels, cross-pieces, and finish up with the vertical side-pieces in the rotation indicated, and a much neater job will result. In this case the hairstrokes follow the grain of the wood.

Breaking Off.—When working on a saloon-house and about to knock off for the night, avoid leaving a plate half finished; it will probably show an unsightly join next day. Always endeavour to finish at the butt of a plate, and don't just go to the edge of it; go well over and this will make the join up next day much neater and easier, with the added advantage that an extra coat of paint is applied to the edge of the plate itself, which is a place where rust starts very easily.

Overheads.—Overhead work is the bane of a sailor's existence, and when doing it (as in other kinds of painting), it is a common fault to see a sailor with a hand plastered with paint which has run from the brush. This is entirely unnecessary, if only a little paint is taken up on the brush, and it is frequently wiped on the edge of the pot.

Covers and Screens.—Old tarpaulins and boat-covers are invariably used as painting screens. They are laid on wooden decks, rails, seats or anywhere else to catch paint spots.

Even if a large screen is not practicable, it is as well to have a small piece of canvas (about 2 or 3 feet square), on which to stand the pot.

Bay.—Section of bulwark between stanchions.

Fleet.—The particular stretch or length a man is able to reach comfortably is referred to as a "fleet".

Painting Aloft.—When a sailor is in his "chair" and ready to start work, it is always safest to send the paint pot up on the signal halliards. The ordinary paint pot is liable to slop over and make a mess on the deck below, but if stood in a bucket, hung on the "chair", all mess will be avoided, whether painting with a wad or brush.

Another method is sometimes used when a wad is employed. The pot is covered over with a piece of canvas tied over the top. In this a slit is made, through which the hand is dipped. On withdrawing the hand, surplus paint is wiped off through contact with the edges of the canvas. This is also a preventive against slopping, but of the two methods the former is the better. Owing to the danger of dermatitis, paint wads are seldom seen now.

Painting Derricks.—If the derricks are not too high off the deck they will be easily reached by standing on a stool or box, but if high and fairly close together a plank will be necessary. If guys are stretched underneath, the plank can be laid across them, or failing that, across the two derricks.

When wide apart, the best method is to sit astride the derrick with the pot slung over it. If the paint pot is attached to a heavy shackle by a lanyard of suitable length, it will hang alongside the derrick at a height which can be adjusted to suit the painter.

Painting Funnels.—Before commencing to paint a funnel or rig gear, *the engineers should be notified about your intentions*. Failure to do so may result in someone being severely scalded—and it may be yourself.

When funnel boards are rigged, they are bolted together in the form of a square, around the funnel, and this is hoisted with a gantline (usually chain), at each corner; but if bosun's chairs are to be used, a hook block and rope gantline will be required for each.

Soot around the top rim will have to be brushed away with either an ordinary deck-broom or perhaps a wire one. Old, cracked or blistered paint will want smoothing off with a chisel scraper, and any white bands will probably require washing before paint can be put on.

If the funnel is hot, tie sacking around the knees, as they will be pressed close to the iron, and the feet should also receive the same attention to prevent ugly toe marks showing on the surface—especially if the funnel is a light-coloured one.

On the top fleet it may be necessary to use a manhelper to join up with the man on either side, while lower down a guy or end of the gantline may have to be used to keep the chair in its proper "fleet".

Some types of paint dry very quickly on account of heat in the funnel. Such fleets will have to be arranged and stroked off in such a manner as not to look too patchy when dry. With others again the effect is directly opposite; they will turn thin and watery and must be applied sparingly, especially near a lower colour which will have to be cut into it.

Funnel Stays.—Stays may be let go and painted at the same time as the funnel itself, while hanging loose, or they can be left until next day, when they may be done in the same manner as other stays set up in place.

Painting Topsides.—Down as far as the waterline is generally known as the topsides, and on this job stages have to be used unless the work is done by shore workers, in which case painting-flats with high trestles are often employed.

As with other surfaces, all loose rust must be scaled off, and all dirt underneath scupper or soil-pipes must be scraped away before any paint is put on. For this reason, two chisel scrapers are always kept handy on the stage.

On the bottom fleet the paint must be cut to the edge of the waterline. Here it meets the boot-topping which, of course, is a different colour.

Even if the waterline is to be cut in later, the paint should be put on sparingly to prevent it "running".

Spinning Pots.—Paint pots used over the side have to be hung on a long lanyard, with the result that, in addition to slopping paint about, they often unlay or chafe through, and the pot and brush is lost. To prevent this it is better to use old plaited log-line.

Marking Stage Ropes.—In order to avoid lowering a stage too far, the stage rope should be marked with a yarn at a height which can be conveniently reached.

Painting Boot-Topping.—Scrubbing or painting boot-topping is mostly done from a boat or painting-flat, and this is kept alongside by means of lines running well forward and aft. These must be made fast on the outside of the boat, or flat, to keep her close to the ship's side.

The special paint used on this work is always a quick-drying composition which requires frequent stirring to prevent it settling.

Much loose rust may be encountered along the waterline; in fact it is always one of the worst places on any vessel and requires constant attention with a scraper.

Care should be taken to cut the topside colour neatly; nothing looks worse than a wavy waterline.

Actually, the part known as boot-topping is really a wide band, extending from the load-line proper to the light load-line. Below that, two different types of paint—anti-corrosive and anti-fouling—are used, but these are only applied in dry-dock and are seldom handled by seafarers.

Scrubbing Boot-topping.—As a vessel rises in the water through the discharge of cargo, and the boot-topping comes into view, it is usually found to be covered with seaweed and barnacles. If the intention is to paint the area exposed, all marine growth must be removed. For this purpose long-handled scrapers and paint scrubbers are used, but if procurable, steel brooms are more effective.

Painting Bow Fleets.—Painting the midship-section of a ship's hull is easy enough—the stage lies close to the flat side, but the bow fleets are more awkward on account of the "flare" or overhang, which is similar to a vertical wedge. This is more pronounced in some vessels than others, and can only be overcome by "bowsing in". In some cases heaving lines are sufficient, but very often something heavier is required. One line is led forward round the stem and made fast on the opposite bow, and another one is led well aft. When the lines are tightened up—either from the stage itself, or by heaving with a winch on deck—the stage will come in close to the bow and lie comparatively steady.

Sometimes it is possible to pass lashings from a porthole to a stage for the same purpose.

Painting Anchor Fleets.—Continually dropping and heaving up the anchor leaves the "fleet" under the hawse-pipe in a very shabby condition. For this reason it is usually the last part to be painted. In addition, the "anchor fleets" are usually "touched up" at the last port, before arrival in the United Kingdom.

Painting Under the Quarter.—Painting under the quarter is even more difficult than the "bow fleets", especially in the case of a vessel with a long counter.

Each ship is a different problem, but as a rule, two or more stages working in conjunction are necessary. Considerable "bowsing in", with lines led to some convenient point is the usual method of getting under the counter, but a long ladder may also be used with advantage if stretched across from one stage to another.

Parts which cannot be reached from a stage are finished off with a boat or painting flat.

Painting Davits.—As a rule the top of a davit can be reached by standing on the lower block, but if not, a stirrup should be rigged between the parts of the fall. A small ladder is even better if one can be found.

In the case of Welin davits, it is better to screw them out a few inches so as to make sure of reaching the parts which are only exposed when the davit is in that position.

Painting Lifeboats.—Two men are usually sent to work together on the job, and they should begin by painting the tops of the davits if they are to be done.

The boat itself is best started on the outboard side amidships, each man beginning on the keel, reaching well out on the bilge and working towards the ends, taking in the boat-chocks as he goes along. When the side of the boat is completed, carry on with the lower part of the davit, then shift over to the inboard side of the boat and start amidships again.

Be sure to spread an old piece of canvas under the boat. Spots on the deck in such a position are not only unsightly, but difficult to reach with a scraper. The cover will also be an advantage if the deck is wet.

Note.—When working outside a boat, never trust your weight to the grab lines around the gunwale. They may be old and rotten.

Painting Steel Decks.—Steel decks require a good deal of scraping, and a good wash down before any paint is applied.

Accumulations of oil around steam winches must also be removed and the waterways dried out. Quick-drying composition paint is invariably used for steel decks.

Painting Winches.—Before paint is applied to a winch, the accumulations of oil and grease must be scraped off, and all parts washed over with paraffin. Occasionally, dark green paint is used, but black is the most popular colour for winches. When not too dirty, a quicker method of cleaning a winch is to use a bucket of caustic, a "turk's head" and the hose.

Painting Galvanised Steel.—No coat of paint will adhere to new galvanising until it has been weathered for about six months or so.

Painting Over Bitumastic.—Paint put on over bitumastic (or *vice versa*), will turn green. The only thing to do is to scrape off the bitumastic, or the paint, whichever has been put on first.

Painting Canvas.—New canvas should be well soaked with water before it is painted. The water shrinks the canvas, and when the paint dries it is held in the shrunken condition, thereby leaving it tightly stretched when dry. A little more turpentine than usual should be used in the paint.

Painting Wires.—A wad may be found to be the best and quickest method of painting a ridge wire but a brush should always be used. *See* "Painting Aloft".

Painting Fish Plates.—When washing or painting fish plates, a "hook chair" is hung on the top edge of the plate. In this the painter sits with his feet braced on the rail of the deck below. The "chair" takes the weight of the body, leaving the hands free for working.

The bucket or paint pot is also hung on the fish plate by means of a pot hook.

Painting Hot Pipes.—Clean off all grease spots. When the pipes are hot, coat over with a metallic paint such as aluminium. This is the only type of paint which will not cause an unpleasant smell when the pipe is hot.

The principal advantage is that it will not discolour or peel off.

Painting Load Lines.—When painting a fleet which includes the load lines, they are usually painted in at the same time, but this means cutting in two colours together which is always a ticklish operation. A better and quicker method is to coat everything over with the topside colour, using a dry brush, then wipe over the markings with a rag damped with turpentine.

When dried off the markings can be filled in more easily with a drop of paint thickened for the purpose.

The proper method is to cut in the topside colour properly, then fill in the markings at a later date, but there is seldom time for this in port.

Painting Draught Figures.—The figures on the stem and stern-post, which indicate the ship's draught, are painted in the same manner as the load lines. A boat is nearly always used on this job, but sometimes a gantline and bosun's chair is rigged for the stem figures.

Painting Lines for Deck Games.—While a drop of thick white or other colour paint is an advantage when painting lines and figures over the side, it is not of much use for lining off deck games. Thick paint will skin over and soon get trodden or scrubbed off. Instead, the paint should be made thin, so that it will soak into the grain of the wood. A lining tool and a piece of wood acting as a straight-edge is the best means of assuring neat, accurate work.

VARNISH AND ENAMEL.

When applied thinly and evenly to any surface, varnish will leave a good, hard, durable coat when a part of the constituents have evaporated. The coat is a good protection from the effects of the atmosphere, and being of a lustrous nature, adds greatly to the beauty of any object or surface.

Best White Varnish.—1 gal. rectified spirits of wine, 2¼ lbs. gum sandarac, ½ lb. gum mastic, 2 lbs. gum anima. Bottle and put in warm place. Shake often and strain before using.

Black Varnish.—Probably more black varnish is used on board ship than any other type, and although there is a large range of black varnishes, the essential characteristic of them all is asphalt and bitumen. To this is added linseed oil, as well as driers consisting of red-lead, litharge and manganese dioxide. The mixture is heated to a high temperature, and when cooled off turpentine is used to bring it to a suitable consistency. Coachbuilders' black Japan is the best grade of black varnish made.

Coloured Varnish.—Sealing-wax dissolved with spirits of wine· Bottle and cork closely.

Copal Varnish.—1 quart spirits of wine, 1 oz. gum copal and ½ oz. shellac. Bottle and keep in warm place. Shake contents until dissolved.

Making Varnish.—Although seamen are not expected to be able to make varnish, a knowledge of how to do so may prove of value. Some of the simpler kinds are as follows:

Oak Varnish.—1¾ lb. pale clear rosin and ½ gal. turpentine.

Polyurethene Varnish.—Gives a very tough, durable coating. Its drying time makes it useful at sea where older varnishes could only be used in the afternoon before about four o'clock on most days, and not at all if the atmosphere was damp.

Spar Varnish.—To 4 lbs. of boiled linseed oil add ½ lb. of best crushed rosin. Stir until dissolved, then add ½ lb. turpentine.

Spirit Varnish.—Is composed mostly of resin dissolved in methylated spirits, turpentine or other volatile spirit. Sometimes a little oil is added to make it more elastic and durable. It is also known as knotting, and is used extensively for covering knots in wood before painting.

Preparing Surfaces for Varnishing.—A good, smooth finish is essential before applying varnish, otherwise it would be a waste of material to coat over a rough surface. If the wood has become dark through successive coats, or is cracked and very rough, it will probably be best to clean off the old varnish right down to the wood. But if not, a rub with pumice-stone (after wetting the surface), or the use of sandpaper will usually be sufficient.

Cleaning off Varnish.—A number of patent preparations on the market are sold for this purpose, but the old-fashioned method of employing caustic soda is still largely used on most vessels. With a small mop improvised from old bagging or rope-yarns, the caustic is laid over the surface and given time in which to bite in and, to help this softening process, the area should be kept wet until ready to scrape away the thickest of the varnish with a scraper.

After another good soaking, a scrubbing-brush and sand will remove the last of the varnish more effectively than the scraper. When finished, wash everything over with salt water, and finish off with fresh.

Care must be taken with the strength of the solution used. If too strong it will burn the wood and give it a brownish tinge which cannot be removed. This scraping and scrubbing process will naturally leave the surface in a very rough state, so it must be finished off with the pumice-stone and sandpaper before varnishing.

It must be remembered, too, that caustic will burn the flesh, therefore, care must be taken to protect the eyes. For this reason a bucket of salt water is always kept handy for such emergencies. Salt water will kill the action of caustic immediately.

Puttying and Filling.—All puttying and filling must be done before any varnish is put on and should be tinted a brownish shade to coincide with the colour of the finished surface.

Oiling Wood for Varnishing.—New wood, or that which has been scraped off clean, is usually oiled first, to act as a filling for the grain of the wood.

Raw oil with a dash of turpentine is used and is put on with a wad as dry as possible. Boiled oil is no good for this purpose as it dries very rough and crinkly.

Applying Varnish.—On board ship varnish is only applied to woodwork. It is never used on steel except in the case of a few vessels where deck houses are grained.

To get the best results, varnish must be applied uniformly. Like enamel, it will form into ridges if put on too thickly, and will lose its gloss if applied too sparingly.

Damp or cold currents of air, as well as a damp surface, will cause varnish to "bloom" or turn a creamy colour; therefore, a dry day when little dust is blowing about is naturally the best for this type of work.

When varnish becomes too thick it is usually through being improperly corked, or the effect of cold weather, but this may be remedied by heating on the galley stove, and standing in a bucket of hot water while in use. As a matter of fact, all varnish should be applied while hot, with the result that it will cover a larger area, be easier to work, give off a better gloss and dry quicker.

Adding boiled or raw linseed oil, turpentine or driers, will spoil the gloss and durability of a varnished surface, but in spite of this, one part turpentine to two parts of varnish is sometimes used for the sake of fast drying.

A brush with plenty of spring is the best for this type of work, but too much varnish should not be taken up on the brush at one time or this will lead to excessive working and cause the varnish to "froth", with the result that the finished surface will be spoiled.

Enamelling.—Enamel should only be applied to a hard, dull surface or over a coat of "flatting". If the surface has been previously enamelled or painted, it must be rubbed down with a pumice-stone or sandpaper.

This is done to remove the gloss and leave a dull surface.

Enamel over new paint containing oil would crack in a few weeks, for the simple reason that oil paints take longer to dry properly than enamel.

As the oil paint surface shrinks through the drying process, the hard enamel coat is disturbed and quickly cracks, which is a thing it will not do if put on over a good foundation.

Turpentine is the only mixture which can be added to make enamel work easier, but should not be used unless absolutely necessary. In cold weather it is better to stand the pot in a warm place for some time, and keep it thin by standing in a bucket of hot water.

Brushes used for enamel should be clean and dry, and the surface to be coated must be free from all dust if the work is to be a success.

Enamel must be applied as uniformly as possible. If put on too thickly, objectionable ridges and patches will form, and if put on too sparingly there will be no gloss.

Cutting In.—This is the art of keeping to a straight line when two colours adjoin. It requires a steady hand and a certain amount of practice to attain any degree of proficiency. A line is always drawn as a guide, and might be either a chalk line or a light scratch made with the aid of a piece of wood acting as a straight-edge. Mostly, however, the edge of the old coat of paint is the only guide, but in time it will become unreliable on account of slight deviations, which must be corrected if the cutting in is to look neat. A fairly well-worn brush appears to be favoured by most men for this work, although it is purely a matter of choice. One good method is to take up plenty of paint on the brush and run it along close to the guiding line for about a couple of feet. Go over it again, but allow the bristles to spread so that they just touch the line and no more as they are drawn along. Cutting in to a curved line is more difficult and calls for a certain amount of judgment to maintain the necessary sweep.

Cutting in a Waterline.—In course of time the waterline will become very wavy and require straightening. This can be done with the aid of a chalk line if the marking of the true waterline can be found. Look down the edges of the plates near the existing line and find the small punch-mark which indicates the true line. They may be rusted over or obliterated altogether, but if found, a line struck from one mark to another will give the true waterline. Failing this, measurements must be taken from the ship's plan and measured off at intervals from the deck to the required distance. When a new line is to be struck, the topside colour is taken well down over the boot-topping in case the old line has crept high in places, which it often does through carelessness.

Mixing Whiting.—The ordinary combination of whiting and water may be good enough for some purposes, but for others a fast surface may be required. This is best done by using a binder such as soft soap or size. About $\frac{1}{4}$ lb. of size dissolved in 1 pint of hot water is usually sufficient for about 3 lbs. of whiting, but more size must be added if a very fast or hard surface is desired. Whiting is simply chalk ground and purified.

Cement Wash.—Cemented waterways and winch beds are often coated over with a weak solution of cement and water. This freshens the surface and gives the impression of being newly cemented.

Cement Boxes.—Small leaks in a vessel's hull can be stopped with the aid of a cement box. This is merely a rough, wooden box built around the place where the leak is situated and filled with cement. When the cement sets hard it is a most effective remedy.

Size.—A weak glue made from isinglass or gelatine, to which is added a large proportion of chondrin, to make it less firm and adhesive. It is one of the constituents of distemper. Ordinary cake glue would answer the same purpose if nothing else was available, but as it takes longer to dissolve it is not quite so convenient.

Cutting Stencils.—Good, stiff paper (such as a piece of old chart) is the best material from which to make a stencil. When the figures or letters are drawn, they should be cut out on a hard surface such as glass, so as to leave a good, clean cut. Once the edges of the letters become moistened with paint they will fray out and lose their shape. To avoid this the whole stencil is coated with patent knotting or a good, hard varnish, and this will leave the paper very hard and stiff and render it impervious to moisture.

Stencil Paint.—A mixture containing half flat and half oil is the best for filling in stencils.

Distemper.—A water paint usually only found on board large passenger vessels. It is used mostly on canvas covers around passenger decks so as to give a snow-white appearance, or it may be used in lieu of white paint in places which always look the worse for wear and tear, and which would not look much better if continually washed.

Distemper is supplied in paste form, in drums, and only requires the addition of hot water when mixing. Should it thicken up when in use, more hot water can be added until it is of the proper consistency.

Emulsion Paint.—In recent years an emulsion paint has been developed which has almost entirely superseded distemper. It contains vinyl acetate which is a tough rubbery plastic and can be thinned out with fresh water. It is easy to apply, dries quickly and sets hard to form a skin which will stand up to repeated washing. Brushes, rollers or containers should be washed out immediately after use and can be thoroughly cleaned in water. It can be applied to most surfaces providing they are clean and will take a second coat within an hour or two of the first. It is available in a large variety of colours.

CHAPTER IX.

MISCELLANEOUS

Anchor Work. Awnings.

ANCHOR WORK.

Anchor and Cable Work.—The proper scope of cable for a vessel at anchor is considered to be at least three times the depth of water. A good scope with a horizontal pull will give good holding power, but a short upward pull is of little use.

Anchoring Procedure.—The anchors are cleared away an hour or so before reaching the anchorage, power is put on the forecastle and the windlass is turned over out of gear. The windlass is then put into gear and each anchor is walked back to loosen it in the hawse pipe; sometimes it is walked back to the waterline before the brake is put on and the windlass is again taken out of gear.

As the ship comes to a stop at the anchorage position, the brake is released and the anchor is dropped to the bottom of the sea; as the ship moves astern, the cable is paid out until at the appropriate place the brake is put on and held until the ship is brought up.

Anchor Balls.—As soon as the anchor is "let go" a black ball 2 feet in diameter is hoisted on the forestay. This indicates the vessel is at anchor. (*See* page 200.)

Clearing a Turn in the Anchor.—Anchors must lie at a certain angle before they will heave up into the hawse-pipe properly. Sometimes even a half turn in the cable will cause them to lie over the wrong way. To take out this half turn, the eye of a mooring-wire is dropped over one of the flukes on the anchor, and is made fast at an angle where it will be most effective (usually at the break of the forecastle head). The windlass is then "walked back", and as the anchor is lowered a weight comes on the wire, and this will exert the necessary pull to twist the anchor in the desired direction. The operation may have to be repeated a few times before it is successful.

Hanging Off an Anchor.—In some ports it is the practice to moor steamers to a buoy with their anchor cable, but before this can be done the anchor must be "hung off" and the cable unshackled. To do this the anchor is lowered until it hangs just clear of the hawse-pipe. A wire is next passed through the shackle of the anchor, led over the warping chocks and up to the mooring bitts. Sufficient turns (in like manner) to hold the anchor should be taken, then hauled tight and made fast. When ready, the windlass is "walked back" until the wire holds the weight of the anchor, and with the windlass still "walking back", the cable is then dragged to one side of the deck as it comes off the fore side of the gypsy, instead

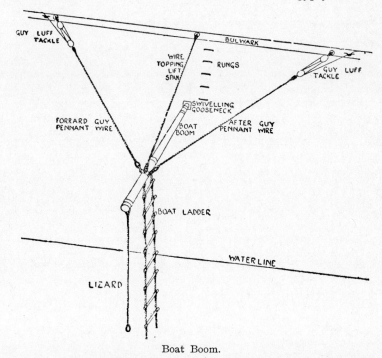

Boat Boom.

of passing down the hawse-pipe. The cable is flaked up and down the deck until the first shackle (15 fathoms) is reached. The pin in this shackle is punched out, and when the cable is disconnected, the free end from the windlass is passed down the hawse-pipe and hove to the buoy with a messenger. The messenger is usually a mooring-wire rove through the ring in the buoy and led back to be shackled on the cable. By heaving away on the wire, the cable is hove out to the buoy and shackled to it.

Clearing Foul Cable.—When lying with two anchors down, a vessel may continually swing the same way round each time the tide changes, with the result that one cable will be twisted around the other. It will then be impossible to heave either one up. Failing the services of a tug-boat (to turn the vessel in the opposite direction) the only thing to do is to either heave or lower the nearest cable shackle to a position just outside the pipe, so that it can be unshackled. By passing a wire around the other cable, in the opposite direction to that which the turns have taken, and by making the same number of turns with the wire, and then shackling it to the disconnected cable, the twists are thus formed into a bight which will free itself when a strain is put on the cables.

As soon as the cables have shaken clear, the disconnected ends are shackled together again, and the anchors are ready for heaving up.

Dragging.—If the anchor drags, the cable is the most likely indicator for the officer of the watch, its movement will be irregular and vibrations will be felt and heard as the cable and anchor drag over obstructions on the sea-bed. Dragging can be stopped by either paying out more cable or by dropping the other anchor as well.

Chain Lockers.—Situated immediately below the windlass are the two compartments where the anchor cables are stowed. Access is gained by a hatch opening directly into it, or through the forepeak.

If they are built with sufficient depth they will be what is known as self-stowing lockers and will require little attention, but if shallow and wide the cable will require to be stowed or it will not all fit in. Actually, the cable is supposed to be flaked down neatly, but there is never time nor sufficient hands to do this properly, so the cable is allowed to pile in anywhere. The usual practice is to let it pile up until the top weight causes it to capsize, but in doing this the cable is liable to become "kinked". This will not be found out until the anchor is let go again, when it may jam in the spurling-pipe, and perhaps put the ship in danger. The safest plan is to guide the cable into the vacant spaces so that the locker fills up evenly. Chain hooks are nearly always provided for handling cable, but are so liable to jam in the links that most sailors discard them altogether and use their hands.

Never go down a locker without a light, or before you get instructions to do so.

The carpenter who works the windlass always shouts his instructions down the spurling-pipe, and they should be repeated and obeyed promptly, especially the order "come out of the locker".

Marking Anchor Hove Up.—In addition to the usual marks on a cable, either a paint- or wire-mark is sometimes placed on a convenient link to coincide with another mark on a stationary part of the windlass. The marks are fixed when the anchor is hove right up in the 'pipe', in order that the carpenter will know the precise moment when steam should be shut off.

Report Loss of Chain-Hooks.—It is not unusual for a chain-hook to become wedged in a link and fall down among the cable in the locker. If this should happen, report the fact to the carpenter.

When the anchor is "let go" again the hook might come up through the spurling-pipe and injure him.

Report Lashed Cables.—To lash both cables together in the chain locker is quite a usual practice in order to prevent them banging about in the spurling-pipe while at sea.

If this is done, the carpenter should be notified in case the lashing is forgotten when entering port.

To be on the safe side, only a very light lashing should be used, one which would break if the anchor was "let go" in an emergency.

AWNINGS.

Spreading Awnings.—When spreading awnings it is sometimes rather difficult at first to tell which way they should be stretched, but by keeping the following few points in mind the task will be greatly simplified.

With an old awning it is easy to find which part is the top and which is the bottom. The top is always discoloured or faded from the effects of the sun. If the awning is a new one, just look for the backbone; it is always on the bottom. The roping is also on the bottom.

Next, to determine which part will stretch forward or aft, examine the seams of the cloths which run athwartships, and you will find two rows of stitching. The row holding down the selvedge edge will be the after end, as the cloths are made to overlap towards aft, so as not to catch the wind and rain which mostly comes from ahead.

When the awning has been roughly spread over the spars, the backbone should be stretched tight and made fast, then the earrings are taken out and made well fast, at the same time taking care that the backbone is not disarranged. After this the side-stops are hitched, and all lacings rove and hauled tight.

Lacing Awnings.—If the lacing is a very long one, the centre of it should be cow-hitched to an eyelet hole about halfway along the side to be laced. The two ends are then worked away from the centre towards the ends.

Making Up Awnings.—Awnings are usually "made up on the backbone". That is, they are folded over double, so that the backbone forms a bight, and the ends and sides are folded in to make the awning a handy size for rolling up. Rolling an awning into a long roll makes it easier to carry and stow away. A few of the stops are always left out in such a manner that they can be used for hitching around the awning, and, if possible, the name which is stamped on the canvas should be left exposed on the outside of the roll for identification purposes. If this cannot be contrived, a canvas label should be attached.

Parts of an Awning.

Backbone.—Wherever there is a collection of awning spars, the central or midship one running fore-and-aft will always be found to be heavier than the others. It bears the most weight, and that part of the awning which rides on it will have to stand considerable chafing.

For this reason, an extra piece of canvas about 1 foot wide is sewn the full length of the awning to take the wear and tear, and being the central and strongest part, it is naturally known as the backbone.

Cloths.—The strips of canvas of which an awning is composed are known as cloths.

Roping.—Rope sewed to the underneath edges or tabling of an awning to strengthen it. The "stops" are spliced into it, thereby taking much of the weight from the canvas, or they may be spliced to eyelets riveted into the canvas.

Earings.—The four corners of an awning are known as earings and are fitted with substantial thimbles and long "stops". These are longer than the "stops" or lashings found at the sides of the awnings. Practically all outstanding corners are known as earings.

Shark's Mouth.—If an awning is cut to allow it to fit neatly around a stay or ventilator, the resulting V cut (or any other shaped cut) is known as a "shark's mouth" or "jaw".

CHAPTER X.

FLAGS

Flag Etiquette. Signalling.

Royal Standard.—The chief flag of the British Commonwealth. It is broken out at the maintop immediately the Queen steps on board.

Origin of Ensigns.—At one time the British Navy was formed into three squadrons—each with a distinctive coloured ensign. These were red, white and blue, but it led to confusion in action, so the practice was discontinued.

White Ensign.—On Nelson's instructions, Trafalgar was fought under one colour—the White Ensign, and this colour has continued as the Naval Ensign since 1846. The Royal Navy and Royal Yacht Squadron have had the exclusive right to wear the White Ensign since 1864.

Red Ensign.—The Red Ensign is now only seen on merchant vessels although some of the British Colonies have also adopted it as their official flag by the introduction of a badge in the fly.

It is interesting to note that vessels of the Hudson's Bay Company use the Red Ensign with the letters H.B.C. in the fly, but they are the only commercial company permitted to add to the ensign in this way.

Blue Ensign.—The Blue Ensign is worn by Government Departments, each one having a distinctive badge in the fly.

It is also worn by merchant vessels commanded and manned by a certain proportion of Naval Reserve Officers and Ratings.

Wearing the Ensign.—Ensigns are only worn on a gaff at sea and on the ensign staff in port. They are always changed over at the moment of arrival or departure.

Definitions.—A ship is said to wear her colours, i.e. the Ensign. A ship is said to fly a personal standard, a distinguishing flag or house flag, i.e. a person flies his (or her) standard or flag "in" a ship or "on" shore.

Dipping the Ensign.—When one vessel wishes to salute another she "dips her ensign", and this is always acknowledged in the same manner.

All ships must "dip" to a naval vessel, and although the procedure differs in various ships (according to the fancy of the master), the following method is perhaps as good as any:

When the bows of the two vessels are nearly in line the ensigns are lowered slowly, and are down when the bridges are abreast. After a short pause they are slowly hoisted.

The proper moment at which to lower or hoist is always indicated by the officer of the watch, either with a whistle or the docking telegraph.

There is no obligation for a merchant vessel to "dip" when passing a vessel of the Royal Yacht Squadron, and the Blue Ensign is not dipped to by the Red.

House Flag.—Private colours or flag of a shipping company. Its proper place is on the mainmast.

Courtesy Flag.—Very often, when leaving port, the national flag of the country to which the vessel is bound is flown on the foremast. Likewise, when lying in a foreign port, the national flag of that country is flown on the same mast as an act of courtesy.

Mail Pennant.—A white pennant with a crown, and the words "Royal Mail" running through the centre in red. It is flown on the foremast by vessels carrying mails.

Flags.—The length of Union Flags, Standards and Ensigns is twice the breadth.

Bunting is supplied in two widths—broad, 19 inches, and narrow, $9\frac{1}{2}$ inches.

To express the size of a flag the term "breadths" is used, this being 9 inches. A flag of 16 breadths would be 12 feet broad and, the length being twice the breadth, 24 feet long.

Note.—The extra half inch will be taken up in the seam.

Bunting of Government make has a few thicker threads worked in every 6 inches of its warp and also along its edges.

Flags (Church Service).—While a service is being held at sea the ensign is always worn and the appropriate courtesy flag, according to the nationality of the passengers, is hoisted.

Jackstaff Flag (or Stem Jack).—The small flag worn on the jackstaff forward has no special significance and is merely ornamental. Some companies use a small house flag but in order to bring about uniformity in the Merchant Navy it has been recommended that all merchant vessels should wear a small pilot jack (a union flag with a white border).

Breaking Out Flags.—Flags are only flown from 8 a.m. till sunset.

Shortly before that time they are sent aloft "in a ball". That is, they are rolled up and made fast in such a manner that they can be quickly released. A hand (usually a boy) is stationed at the halliards of each flag just before 8 a.m. When he hears the bell he jerks the downhaul part and the flag flies clear. In this manner they are all "broken out" together.

The ensign is hoisted in the ordinary way and is never "broken out".

INTERNATIONAL CODE OF SIGNALS

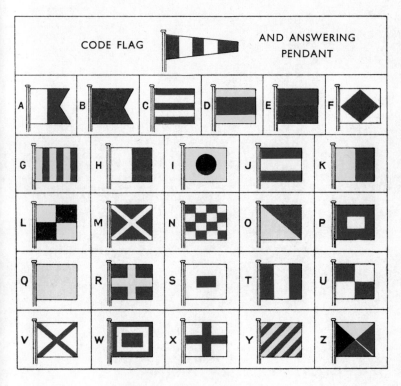

NUMERAL PENDANTS.

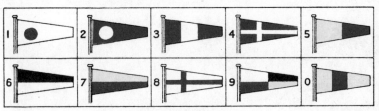

SUBSTITUTES.

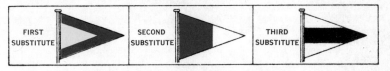

Sunset.—When flags are hauled down at sunset, the ensign should come down last, and if a Naval vessel is in the vicinity, all merchant ships "take the time" from her. That is, they "break out" and haul down their flags at the same time as the naval ship.

Flags Half-Mast.—Masthead flags, such as the house flag, should only be half-masted in mourning for owners, partners in the firm, or members of their families. Other masthead flags, like mail pennants and foreign ensigns, should never be half-masted.

Dressing Ship.—On special occasions a ship is dressed by bending sufficient flags together to stretch from the stem head to the fore and main truck, then down to the poop. This is known as "rainbow style", and any order of flags may be used. Owing to the weight of the flags, masthead gantlines are used for hoisting them up, and the flags are usually bent to a wire for the full distance aloft.

It has been pointed out that "rainbow fashion" is a misnomer when used in connection with "dressing ship". Over one hundred years ago the mast profile of most vessels resembled an "arch", and this is generally supposed to have been the origin of the term, but by no stretch of imagination could the present-day profiles described as arches or likened to rainbows.

However, until someone suggests a better word, the term will probably remain in use.

Decoration Flags.—Large passenger vessels carry a number of extra flags which are used solely for decorating decks when holding dances, etc.

Ships' "Number".—The hoist with four letters indicating a ship's name is generally called her "number", and they are kept bent together, so as to be ready for hoisting when wanted. Each ship is allotted a permanent official number.

Signalling.—Communicating messages by means of flag signals is of very ancient origin, going back as far as Athenian times.

Only very simple messages could be sent in those days, but the development of naval warfare eventually caused codes to be devised, in order that a limited number of flags could, by continually altering their position in a "hoist", convey a greater number and variety of meanings.

At the present time the code consists of 40 flags, which, in addition to expressing some thousands of set phrases by means of small alphabetical groups indicated by flags, can also be used for spelling sentences not so listed. Actually there is no limit to the messages which it may be desired to send.

INTERNATIONAL CODE OF SIGNALS.

The 1969 edition of this publication has been reduced to a single volume and has been very much simplified to give each signal a complete meaning with Complements to allow for possible variations in the basic meaning. It can be used in conjunction with all methods of communication, namely by:—

(a) Flags. (The 40 flags are shown facing page 238).

(b) Flashing light. (Morse.)

(c) Sound signalling. (Whistle, siren, foghorn, etc.)

(d) Voice over a loudhailer.

(e) Radiotelegraphy.

(f) Radiotelephony.

(g) Signalling with hand flags or arms,
    (i) Semaphore, (ii) Morse.

PHONETICS.

It is suggested that the following internationally accepted and uniform system of phonetic alphabet and figure spelling tables should be used when reading or transmitting plain language or code:—

Letter	Word	Pronounced	Letter	Word	Pronounced
A	Alfa	*Al* fah	N	November	No *vem* ber
B	Bravo	*Brah* voh	O	Oscar	*Oss* cah
C	Charlie	*Char* lee	P	Papa	*Pah* pah
D	Delta	*Dell* tah	Q	Quebec	*Key* beck
E	Echo	*Eck* oh	R	Romeo	*Row* me oh
F	Foxtrot	*Foks* trot	S	Sierra	See *air* rah
G	Golf	Golf	T	Tango	*Tang* go
H	Hotel	Hoh *tell*	U	Uniform	*You* nee form
I	India	*In* dee ah	V	Victor	*Vik* tah
J	Juliett	*Jew* lee *ett*	W	Whiskey	*Wiss* key
K	Kilo	*Key* loh	X	X-ray	*Ecks* ray
L	Lima	*Lee* mah	Y	Yankee	*Yang* key
M	Mike	Mike	Z	Zulu	*Zoo* loo

Figure	Word	Pronounced
0	Nadazero	Nah-dah-zay-roh
1	Unaone	Oo-nah-wun
2	Bissotwo	Bees-soh-too
3	Terrathree	Tay-rah-tree
4	Kartefour	Kar-tay-fower
5	Pantafive	Pan-tah-five
6	Soxisix	Sok-see-six
7	Setteseven	Say-tay-seven
8	Oktoeight	Ok-toh-ait
9	Novenine	No-vay-niner

In addition, the Decimal Point is indicated by the word "DECIMAL", the Full Stop by the word "STOP", and any change to Code groups to follow is indicated by the word "INTERCO".

FLAG SIGNALLING PROCEDURE.

The signal letters of the vessel addressed are hoisted with the signal, though as a general rule only one hoist should be flown at a time but when necessary to have other groups on the same halyard, they must be separated by a tackline. The vessel addressed answers by hoisting the answering pendant at the dip and when the signal is understood, by hoisting the pendant close up. This is repeated for each group until the sending vessel hoists the answering pendant close up as a single flag signal to show the message has been completed.

Use of Substitutes.

Three flags—first, second and third substitute—which enable any letter or number of an alphabetical or numerical signal to be repeated, without using more than one set of flags.

The First Substitute.—Can only repeat the top flag of those which may precede it in that particular class and group.

The Second Substitute—Can only repeat the second flag (counting from the top) of those which precede it in that particular class and group.

The Third Substitute.—Can only repeat the third flag (counting from the top) of those which precede it in that particular class and group.

A substitute cannot be used more than once in the same group.

Groups.

The importance and nature of any group of signals can be seen at a glance, this depending on the number of flags in the hoist, as well as other indications, such as the top flag, etc.

One-Flag Signals.—Are urgent, important or of common use and are easily memorised.

Two-Flag Signals.—Are from the General Code.

Three-Flag Signals (with the lowest flag a numeral).—Are also from the General Code.

Three Flag Signals (M uppermost).—Are from the Medical Code.

The following single-letter signals with complements are quite distinctive:—

Flag A with 3 numerals indicates an Azimuth or Bearing.

Flag C with 3 numerals indicates a Course.

Flag D with 2, 4 or 6 numerals indicates a Date.

Flag G with 4 or 5 numerals indicates a Longitude.

Flag L with 4 numerals indicates Latitude.

Flag R with 1 or more numerals indicates Distance in Nautical Miles.

Flag S with 1 or more numerals indicates Speed in knots.

Flag T with 4 numerals indicates Time (local).

Flag V with 1 or more numerals indicates Speed in kilometres per hour.

Flag Z with 4 numerals indicates G.M.T.

Flag K with 1 numeral indicates wish to Communicate (numeral indicates method from Complements Table 1).

Two other useful code signals are:—

(i) YZ "The words which follow are in plain language."

(ii) YV "The groups which follow are from the International Code."

CHAPTER XI.

NAVIGATIONAL EQUIPMENT

Points of the Compass. Steering the Ship. Lead Line and Soundings. Other Instruments and Electronic Equipment.

Compass Cards.—At one time Compass Cards were marked in points, half-points and quarter-points but with the passing of the days of "sail" and the increasing efficiency in the means of steering, all cards became additionally marked in degrees. At first these graduations were from 0 to 90 between the four cardinal points but

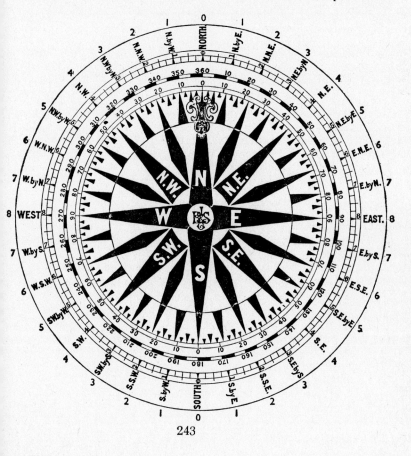

today all cards are marked consecutively from 0 to 360 from North in a clockwise direction. In small craft steering may still be carried out in points and a sailor should be able to "Box the Compass" by knowing and naming the points in their proper order.

Markings on the Card.—Altogether, there are 32 points on the compass card. The most important are the 4 "cardinal points", namely, N., S., E. and W. Next in importance come the 4 "quadrantal points". These are N.E., N.W., S.E. and S.W. They indicate the four respective quadrants of the compass. The names are printed on them.

Then we have 8 "three-letter points". The names are formed by prefixing the name of the nearest cardinal to the quadrantal point. We then get: N.N.E., E.N.E., E.S.E., S.S.E., S.S.W., W.S.W., W.N.W., and N.N.W.

Lastly, we have 16 "by points", four in each quadrant. They are the points immediately to the right and left of the cardinals and quadrantals and take their names from them, as indicated in the adjoining compass card and tabulation. The name "by" means, one point. Thus . . . one point to the right of the north is referred to as, N. by E., and one point to the left of north as N. by W. Similarly, we read N.E. by N., N.E. by E., E. by N., E. by S., and so on round the card.

Learning Quarter Points.—Take note that the three-letter points are used only to indicate their own direction, and that the quarter points are read from the nearest "cardinal" and, "by" point as given in the Table. Note, also, that the "quarters" between the three-letter points are named from the quadrantal point of their respective quadrants. (*See* next page.)

Thus, in splitting into quarter points the direction between two given points we say, for example, N. by E. $\frac{1}{4}$ E., N. by E. $\frac{1}{2}$ E., N. by E. $\frac{3}{4}$ E., then skip N.N.E., and read N.E. by N. $\frac{3}{4}$ N., N.E. by N. $\frac{1}{2}$ N., N.E. by N. $\frac{1}{4}$ N., N.E. by N. and so on.

Lubber's Line.—A clearly visible thin black line is marked inside the compass bowl to indicate the direction of the ship's head and is known as the "Lubber's Line". When the compass is installed the line through the centre of the pivot and the lubber's line is lined up parallel to the fore-and-aft midship line of the ship.

Steering.—Learning to steer depends on individual effort, and proficiency can only be acquired by considerable practice.

Before attempting to learn, the beginner should have a thorough knowledge of the compass card, and should be familiar with the principle of steering, as well as the various orders likely to be given to a helmsman.

The first thing to be remembered is that the top of the wheel must be turned in the direction you want the ship's head to go, or in other words, when a vessel "falls off" her course (as shown by the

QUARTER POINTS.

North	North	South	South
N. $\frac{1}{4}$ E.	N. $\frac{1}{4}$ W.	S. $\frac{1}{4}$ E.	S. $\frac{1}{4}$ W.
N. $\frac{1}{2}$ E.	N. $\frac{1}{2}$ W.	S. $\frac{1}{2}$ E.	S. $\frac{1}{2}$ W.
N. $\frac{3}{4}$ E.	N. $\frac{3}{4}$ W.	S. $\frac{3}{4}$ E.	S. $\frac{3}{4}$ W.
N. by E.	**N. by W.**	**S. by E.**	**S. by W.**
N. by E. $\frac{1}{4}$ E.	N. by W. $\frac{1}{4}$ W.	S. by E. $\frac{1}{4}$ E.	S. by W. $\frac{1}{4}$ W.
N. by E. $\frac{1}{2}$ E.	N. by W. $\frac{1}{2}$ W.	S. by E. $\frac{1}{2}$ E.	S. by W. $\frac{1}{2}$ W.
N. by E. $\frac{3}{4}$ E.	N. by W. $\frac{3}{4}$ W.	S. by E. $\frac{3}{4}$ E.	S. by W. $\frac{3}{4}$ W.
N.N.E.	**N.N.W.**	**S.S.E.**	**S.S.W.**
N.E. by N. $\frac{3}{4}$ N.	N.W. by N. $\frac{3}{4}$ N.	S.E. by S. $\frac{3}{4}$ S.	S.W. by S. $\frac{3}{4}$ S.
N.E. by N. $\frac{1}{2}$ N.	N.W. by N. $\frac{1}{2}$ N.	S.E. by S. $\frac{1}{2}$ S.	S.W. by S. $\frac{1}{2}$ S.
N.E. by N. $\frac{1}{4}$ N.	N.W. by N. $\frac{1}{4}$ N.	S.E. by S. $\frac{1}{4}$ S.	S.W. by S. $\frac{1}{4}$ S.
N.E. by N.	**N.W. by N.**	**S.E. by S.**	**S.W. by S.**
N.E. $\frac{3}{4}$ N.	N.W. $\frac{3}{4}$ N.	S.E. $\frac{3}{4}$ S.	S.W. $\frac{3}{4}$ S.
N.E. $\frac{1}{2}$ N.	N.W. $\frac{1}{2}$ N.	S.E. $\frac{1}{2}$ S.	S.W. $\frac{1}{2}$ S.
N.E. $\frac{1}{4}$ N.	N.W. $\frac{1}{4}$ N.	S.E. $\frac{1}{4}$ S.	S.W. $\frac{1}{4}$ S.
N.E.	**N.W.**	**S.E.**	**S.W.**
N.E. $\frac{1}{4}$ E.	N.W. $\frac{1}{4}$ W.	S.E. $\frac{1}{4}$ E.	S.W. $\frac{1}{4}$ W.
N.E. $\frac{1}{2}$ E.	N.W. $\frac{1}{2}$ W.	S.E. $\frac{1}{2}$ E.	S.W. $\frac{1}{2}$ W.
N.E. $\frac{3}{4}$ E.	N.W. $\frac{3}{4}$ W.	S.E. $\frac{3}{4}$ E.	S.W. $\frac{3}{4}$ W.
N.E by E.	**N.W. by W.**	**S.E. by E.**	**S.W. by W.**
N.E. by E. $\frac{1}{4}$ E.	N.W. by W. $\frac{1}{4}$ W.	S.E. by E. $\frac{1}{4}$ E.	S.W. by W. $\frac{1}{4}$ W.
N.E. by E. $\frac{1}{2}$ E.	N.W. by W. $\frac{1}{2}$ W.	S.E. by E. $\frac{1}{2}$ E.	S.W. by W. $\frac{1}{2}$ W.
N.E. by E. $\frac{3}{4}$ E.	N.W. by W. $\frac{3}{4}$ W.	S.E. by E. $\frac{3}{4}$ E.	S.W. by W. $\frac{3}{4}$ W.
E.N.E.	**W.N.W.**	**E.S.E.**	**W.S.W.**
E. by N. $\frac{3}{4}$ N.	W. by N. $\frac{3}{4}$ N.	E. by S. $\frac{3}{4}$ S.	W. by S. $\frac{3}{4}$ S.
E. by N. $\frac{1}{2}$ N.	W. by N. $\frac{1}{2}$ N.	E. by S. $\frac{1}{2}$ S.	W. by S. $\frac{1}{2}$ S.
E. by N. $\frac{1}{4}$ N.	W. by N. $\frac{1}{4}$ N.	E. by S. $\frac{1}{4}$ S.	W. by S. $\frac{1}{4}$ S.
E. by N.	**W. by N.**	**E. by S.**	**W. by S.**
E. by $\frac{3}{4}$ N.	W. by $\frac{3}{4}$ N.	E. $\frac{3}{4}$ S.	W. $\frac{3}{4}$ S.
E. $\frac{1}{2}$ N.	W. $\frac{1}{2}$ N.	E. $\frac{1}{2}$ S.	W. $\frac{1}{2}$ S.
E. $\frac{1}{4}$ N.	W. $\frac{1}{4}$ N.	E. $\frac{1}{4}$ S.	W. $\frac{1}{4}$ S.
East	**West**	**East**	**West**

movement of the lubber's line across the graduations on the compass card) the wheel should be turned in the opposite direction to bring her back again.

Taking the case of a vessel which is "steady" or "right on" her course with the wheel amidships when she starts to "fall off" to starboard, the helmsman would immediately counteract this by turning the wheel in the opposite direction (port or left wheel) to "bring her back" to her course.

When she has stopped "falling off" and is starting to come back, the wheel is again "taken off" and brought back to the "midships".

If she stops moving when on her course all will be well, but if given too much wheel, and it is kept on too long, her head may start to fall off to port. When this is about to happen the helmsman will "meet her" by putting the wheel to starboard for a few moments, then bring it back to the "midships" position again when her head has stopped moving.

The ship is now "steady" on her course, and may remain there a few moments before "falling off" once more. When she starts to do so the same procedure is gone through. Wheel opposite way, "midship", "meet her" (if necessary), then "midships" again.

Steering consists of a repetition of these movements, but the amount of wheel to use, or length of time to keep it on is where the skill comes in. It is a matter of judgment, and this can only be acquired with practice.

An important point to remember is to give the ship as little wheel as possible. The man who steers the straightest course with the least amount of wheel action is the best helmsman. By keeping one eye on the tell-tale or indicator, the helmsman can see the amount of wheel he is using at any particular moment.

Steering Orders.—The most frequent order given to the man at the wheel is "port easy" or "starboard easy". If the alteration in the course is only a small one it will soon be followed by "steady", or "steady as she goes", and the helmsman will then note the position of her head on the compass and steer that course.

Should the alteration be a large one, and the ship is swinging round quite fast and nearing her new course, the order "ease the wheel" will be given, upon which some wheel is "taken off" to slow her down. Next comes the order "midships", and when carried out the ship will then swing very slowly, until "steady" or "steady as she goes" is given.

On the order "hard a port", or "hard a starboard", the wheel is put "hard over", or as far as she will go in the required direction, and this will usually be about three or four turns or even more. Never *jam* the wheel "hard over" or damage may result. When the limit has been reached, ease back a couple of spokes.

When the order is given as "starboard 10", the wheel should be turned to starboard until the helm indicator shows that the rudder is 10 degrees to starboard and similarly if the order is "port 5", the wheel should be turned to port until the indicator rests at 5 degrees to port.

When steering by a landmark, you might be asked "how is her head now" in answer to which you will give the course you happen to be steering, but if you are steering by compass, and the ship is on her proper course, you may reply by saying "right on". If you are not on the proper course, give the direction of the ship's head at that particular moment.

Sometimes in the case of a vessel not steering too well, and close to the land or another ship, you might be told "nothing to the nor'ard", or it may be the southward, eastward or westward, according to the cardinal points between which you are steering. Great care should be taken to observe this order and prevent the vessel swinging in the prohibited direction. An overtaking vessel may be very close to your stern.

Under the same circumstances of a vessel not steering too well, and especially when passing another vessel, the order "steer small" might be given. To do this, use more wheel than usual, but use it quickly, thereby causing the ship to make shorter swings.

When a vessel shows a tendency to swing to one side of her course only, the order "keep her as much on one side as the other" is sometimes given. This ensures the ship making a straighter course.

At least twice each watch, and always after altering course, the officer of the watch will "steady her on". This is done to find any alteration in the difference between the standard and steering compasses. You will probably be informed when this is to take place, so that when the officer sings out "right on" or stamps his foot on the deck overhead, you will be ready to instantly note the direction of the ship's head by the steering compass.

Various circumstances will often require that a certain amount of wheel be kept on all the time. When this happens a vessel is said to be "carrying port helm" or "carrying starboard helm", as the case may be.

All orders given to the man at the wheel must be repeated by him in a loud voice, to make sure that he has heard them correctly, and again when the order has been carried out.

When the trick at the wheel is finished and the helmsman about to be relieved, he must hand the ship over when she is steady on her course, and must report the course clearly to his relief, who must repeat it to make sure that he has heard it correctly. The relief must also be told how much wheel the vessel is taking, and how much helm she is carrying, if any. Before he leaves the wheel, the helmsman should mention whether or not the wheel is amidships at that particular moment. He then reports the course to the officer of the watch to avoid any misunderstanding.

With the helm amidships, a right-handed propeller (when the engines are going ahead) has the effect of driving a vessel's head to port. When going astern her head would go to starboard. The effect is greatly increased when a vessel is "light".

SOUNDINGS.

This refers to the methods of finding the depth of water and includes the use of the hand lead, the deep-sea lead, sounding machine and echo sounding devices. "Taking soundings" is the usual expression but the first of the methods mentioned above is referred to as "taking a cast of the lead".

Hand Lead.—The leadsman stands on a small platform amidships which projects over the side about 2 feet. This platform is known as the "chains". It is fitted with a canvas "apron", against

I

which the leadsman leans while heaving the lead, and which protects his legs from the wet line.

Leads.—The lead itself weighs from 7–8 lbs., but a heavier one of 14 lbs. is preferred by many leadsmen, because it sinks much quicker and better soundings can be taken—especially if there is much way on the vessel.

Markings on Line.—Although the full length of a lead line is 25 fathoms, 5 fathoms of this is drift line, and the other 20 is marked off with 9 marks and 11 deeps. The *marks* are as under:

At	2 fathoms	-	-	-	A piece of leather with 2 tails
,,	3 ,,	-	-	-	A piece of leather with 3 tails
,,	5 ,,	-	-	-	White linen
,,	7 ,,	-	-	-	Red bunting
,,	10 ,,	-	-	-	Piece of leather with hole in it
,,	13 ,,	-	-	-	Blue serge
,,	15 ,,	-	-	-	White linen
,,	17 ,,	-	-	-	Red bunting
,,	20 ,,	-	-	-	Cord with two knots

The *deeps* areas follows:

The fathoms 1, 4, 6, 8, 9, 11, 12, 14, 16, 18 and 19 are known as "deeps", and these are sometimes indicated with a small piece of marline, though it is not a general practice to mark deeps.

The Hydrographer of the Navy has introduced the following metric markings for lead lines, which are still used for survey work (mainly for the regular calibration of echo sounders, for variations in the speed of sound in water).

1, 11 and 21 metres	..	..	..	one strip of leather
2, 12 and 22 metres	..	..	..	two strips of leather
3, 13 and 23 metres	..	..	..	blue bunting
4, 14 and 24 metres	..	..	..	green and white bunting
5, 15 and 25 metres	..	..	..	white bunting
6, 16 and 26 metres	..	..	..	green bunting
7, 17 and 27 metres	..	..	..	red bunting
8, 18 and 28 metres	..	..	..	blue and white bunting
9, 19 and 29 metres	..	..	..	red and white bunting
10 metres	..	..	..	..	leather with a hole in it
20 metres	..	leather with a hole in it and 2 strips of leather
30 metres	..	leather with a hole in it and 3 strips of leather
40 metres	..	leather with a hole in it and 4 strips of leather
50 metres	..	leather with a hole in it and 5 strips of leather

All 0·2-metre markings   ..   ..   .. a piece of mackerel line

One advantage of this system is that the metre markings almost correspond to the fathom marks on the conventional leadline.

Hints for Leadsmen.—On first entering the chains see that the end of the line is made fast, and be sure to coil it left-handed or a multitude of kinks will result. The length of line for swinging depends on individual taste, especially if a to and fro motion only is used. If the lead is to be swung over the head, about two fathoms will be found most convenient. It is usual to swing three times over the head, but if the water is deep and the ship travelling fast, a longer drift will be advisable.

To prevent the line slipping through the hand a small bight is placed over the thumb, but this will soon take the skin off, so a small flag toggle should be fitted to the line and used instead.

Calling the Soundings.—The actual number of fathoms must be made the last part of the call, as it is quite likely to be heard even if the first part of the call is missed. Call them as under:

At a depth of 5 fathoms call	-	-	by the mark 5
,, $5\frac{1}{4}$,,	-	-	and a quarter 5
,, $5\frac{1}{2}$,,	-	-	and a half 5
,, $5\frac{3}{4}$,,	-	-	a quarter less 6
,, 6 ,,	-	-	by the deep 6
,, $6\frac{1}{4}$,,	-	-	and a quarter 6
,, $6\frac{1}{2}$,,	-	-	and a half 6
,, $6\frac{3}{4}$,,	-	-	a quarter less 7
,, 7 ,,	-	-	by the mark 7

and so on. If you fail to get bottom, sing out "no bottom at 8 fathoms", or whatever the amount of line out indicated.

Other Uses.—When a vessel is coming to an anchor and the way is off the ship, the leadsman should keep his lead on the bottom, and when the vessel is moving astern, he must immediately report "going astern, sir".

And again, when the vessel is brought up or stationary at her anchor, he should report "brought up, sir".

Sounding at Night.—When sounding at night it is necessary to illuminate the water surface by means of an electric light cluster, searchlight or hand torch. Otherwise readings will have to be taken "from the hand" by first of all finding the distance from the hand to the water's edge. This amount is then deducted from all soundings and because of the different materials used for each colour, the leadsman should be able to know the marks by feel even when no light is available.

Marking a New Line.—Splice an eye in one end large enough to slip over the lead when it has been passed through the grommet. Stretch and soak the line. Instead of measuring with a rule, chalk 3 fathom lengths on the deck, and measure the line by comparing it with the chalked lengths. Attach marks by tucking them through the strands.

Deep-sea Lead.—The lead itself weighs from 28–30 lbs., and the line, which is kept on a small wooden hand reel, is marked in a similar manner to the hand lead, up to 20 fathoms. After this, every 10 fathoms is marked with an additional knot, and every 5 fathoms in between with a single knot.

For instance, at 25 fathoms there is one knot; at 30 fathoms 3 knots; 35 fathoms 1 knot; 40 fathoms 4 knots and so on, up to 100 fathoms which is marked with a piece of leather and two holes.

All leads have cavities at the bottom end so that they may be armed with tallow to ascertain the nature of the sea bottom for comparison with the chart.

Taking a Cast.—Run the line forward (outside of everything) from amidships to the break of the forecastle head, bend on lead, and when ready, notify the man in the chains by singing out "watch there, watch", and drop it over the side. A small wooden snatch block is hung in a convenient position for hauling the line in again.

Arming the Lead.—In the bottom of the lead is a small cavity into which a mixture of white-lead and tallow is tightly pressed. When this strikes the sea bottom, small particles adhere to the mixture and are brought to the surface for comparison with similar data on the chart.

This is a valuable aid to navigation during fog, and a number of soundings will often indicate the course of a vessel fairly accurately.

Benefit of Lead.—As the marks on the leadline are measured from the eye, the length of the lead about 230mm means that there is always a little more water than the mark indicates.

INSTRUMENTS AND EQUIPMENT.

Navigation generally refers to the passage of a craft between two separated positions. Marine navigation is concerned with the passage of a vessel between positions completely interconnected by waterways, whether these be oceans or canals, or a combination of both. It is the duty of those aboard to ensure that the vessel travels between ports in a safe and efficient way. In order that a ship's navigator may perform his duties effectively, he must be provided with sufficient navigational equipment. The amount and variety of apparatus found in a vessel is determined partly by the trade in which she is engaged and partly by the amount of financial investment in her. The equipment described below is to be found in most ordinary ships.

In travelling from one place to another on the earth's surface it is necessary to know in which direction the ship is going at any instant, and her speed or distance travelled, for which purpose the following items of equipment are provided.

COMPASSES.

There are two types of compass differing fundamentally in their mode of operation and can be classified as follows:

    (i) The magnetic compass.   (ii) The gyro compass.

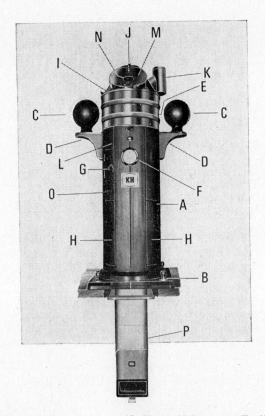

(a) Binnacle Stand.	(i) Hood, Helmet or "Top".
(b) Deck Bolts.	(j) Rifle Sights.
(c) Corrector Spheres.	(k) Fitting for Oil Lamp.
(d) Brackets.	(l) Dimmer Plate Handle.
(e) Binnacle Band.	(m) Night Shades.
(f) Clinometer.	(n) Day Blinds.
(g) Electric Light Chamber.	(o) Flinders Bar. (On the foreside.)
(h) Chamber for Magnets.	(p) Wheelhouse Projection Lens System.

FIG. 1.—The Transmitting Magnetic Compass

Magnetic Compass.—The first mention of a compass is made in an old Chinese legend dating back to 2634 B.C. but it was not until A.D. 1187 that records show it to have been in common use in Northern Europe.

The magnetic compass uses the inherent magnetic forces within and encircling the earth in order to establish direction. A magnetic needle, when freely suspended, is influenced by the earth's magnetic field and aligns itself approximately in a north-south direction. The modern magnetic compass uses the same principle of operation as the earliest Chinese lodestone suspension, but elaborate means are used to give the modern compass freedom from instability and local ship effects.

The Lubber's Line is painted on the inside of the bowl which is both air and watertight and the whole is mounted on gimbals in the binnacle stand. The bowl has a glass base so that the card can be illuminated from underneath and a dimming device is fitted to assist when taking bearings at night.

To counteract any local magnetic influence the binnacle is made to accommodate corrector magnets, athwartship soft iron spheres and a Flinder's Bar consisting of short lengths of cylindrical vertical soft iron contained in a brass case on the foreside of the stand.

A "clinometer" is usually fitted on the afterside of the binnacle to indicate the number of degrees of heel or the extent to which the vessel may be rolling.

Gyro Compass.—The gyro compass uses the properties of gyroscopic inertia and precession. A child's conical top, when not spinning, will topple over. If it is made to rotate rapidly it will not deviate from the upright position. This is an example of gyroscopic inertia., i.e. the axle of a rotating body tends to remain pointing in a fixed direction.

If a force could be applied to the rotating parts of a spinning body in such a way that it were not slowed down, it would be seen that the object would move in a direction contrary to that expected. This is the property of precession exhibited only by rotating objects. These two properties of rotating bodies are harnessed in the gyro compass to produce a mechanism which continuously points to north, provided its sensitive rotating parts are kept energised from a suitable electrical power supply.

The magnetic north and south direction defined by the magnetic compass is not the true geographical north and south directions required for the purposes of navigation. Corrections are either tabulated or charted for all positions on the earth's surface, so that when applied to the magnetic compass card, the true direction of north and south may be readily found.

Although the gyro compass ideally seeks to align itself in the true north-south direction, it is like most mechanical apparatus:

subject to small errors, which must be allowed for when the compass is in use.

Liquid Compass.—In small craft, where rolling and pitching is more pronounced than in larger vessels, a "dry" magnetic compass would be of little use, so liquid magnetic compasses containing a mixture of water and alcohol are used. The effect of the liquid

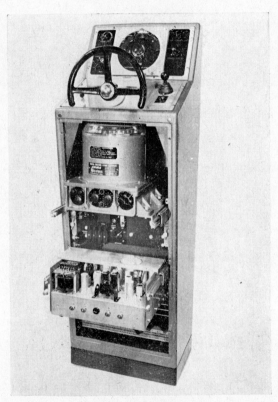

Fig. 2.—Auto Electric Steering Control Console.
(Arma Brown Gyro Compass.)

upon the surface of the compass card is to cause it to remain steadier and settle quicker than would be the case if the liquid were omitted.

To allow for the expansion or contraction of the liquid in hot or cold climates, the compass bowl is fitted with expansion chambers. By additional damping devices, liquid compasses are today mostly of the "Dead Beat" type in which oscillations are almost completely eliminated and on deflection the card returns to its original position in one swing without overswinging.

Lifeboat compasses are all of the liquid type.

Standard Compass.—Although a vessel may be fitted with gyro compass equipment, she is still obliged to carry a standard magnetic compass in addition to a steering magnetic compass. In some vessels the standard and steering compasses are combined by using a light projector from the standard compass down to the steering position. The requirements for the location of the standard compass in a vessel are such that it can reasonably only be positioned on the "Monkey-island".

Automatic Steering.—A vessel, having passed from congested or coastal waters to the open sea, usually switches over from manual to automatic steering, provided she is so equipped. An immediate advantage is that in most sea conditions in which a vessel is likely to steam, the track kept by the automatic system is an improvement on that which could be maintained by a quartermaster for prolonged periods. Generally, less helm is used to steer the vessel, and this results in higher log speeds being recorded. The difference may amount to one-half of a knot in ordinary sea conditions. Like most modern equipment, there is very little of the "automatic" character of these devices to be seen on the control panel and frequently the only outward sign is a switch indicating "Automatic/ Manual". There are usually a number of controls attractively arranged and Fig. 2 shows such a console with the front panel removed.

LOGS.

To find the speed of a vessel at sea it was once the custom to throw a log of wood over the bow and note the time taken for the vessel to pass it. This time interval was ascertained by repeating a set number of words, the time for saying which was known. When the log passed the stern, the number of words spoken gave a time interval, which in turn indicated the speed of the ship. The use of such logs has given us a name to one of the aids to navigation which is still in use at the present time ; but in place of the old-fashioned system we now have the Patent Log.

Patent Log.—The log is really a type of speedometer which registers the number of miles the ship has run. It is usually connected to the rail right aft, although in some vessels it may be fitted to a boom amidships where it can be read on the bridge.

It is a small cylindrical brass case with a pointer which indicates the number of miles on a dial graduated from 0–100.

The latest and most popular types have a smaller hand which registers tenths of a mile.

Attached to this mechanism on the rail is a long length of special line towing in the water, and revolving through the agency of a rotator at the end. It is these revolutions which keep the mechanism on the rail in motion, and a uniform speed is maintained by the use of a small wheel or governor close to the mechanism itself.

Streaming the Log.—First of all the cylinder or clock is shipped into its proper socket and the governor hooked on, and when the logline itself is hooked to the governor, the rotator is lowered into the water and all the line paid out. Make sure the line is clear for running by flaking it up and down the deck beforehand.

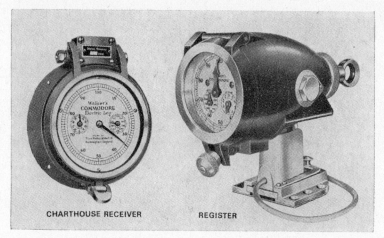

CHARTHOUSE RECEIVER   REGISTER

Fig. 3.—Walker's 'Commodore' Log Parts

Hauling in the Log.—Pull in some slack, unhook from governor, and as the line is pulled in on one quarter, pay out again over the other, so that when pulled in a second time, it will be free from kinks. Coil downleft-handed, hang it up to dry, then put the clock in the chartroom or wheelhouse for safety.

Streaming the Midship Log.—For convenience it has become the practice to stream the log from amidships from a position just abaft the Navigating Bridge and in this case a special boom has to be used to keep the line and rotator clear of the ship's side. When the length of wire which transmits the revolutions from the universal joint connector on the end of the boom to the clock on the bridge is hooked on, and the logline and governor have been attached, the boom is ready for swinging out. With the two guys and the topping lift tackle it can be adjusted to the most suitable position. Either before or after swinging out the boom, stretch the rotator end of the logline down the ship's side (outside of everything), to the break of the poop.

If put over the side from aft it is less likely to be damaged against the ship's hull.

Taking in the Midship Log.—Either a tripping line or small grapnel is used when taking the logline on board again, and the rotator must be hauled carefully up the side to prevent it becoming damaged, after which the boom may be swung in. If the boom was swung in first, the rotator would revolve against the ship's hull and bend the blades.

Speed by Log.—To find the speed at which a vessel is travelling, divide 3600 by the number of seconds taken to run one mile, and this will give the speed in knots.

In the case of a log with a bell which strikes six times to each mile run, divide 600 by the time in seconds between two strokes of the bell, and the answer will be in knots, or miles per hour.

Electrical Recording.—The indicator mechanism of the patent log can be made to generate an electrical voltage which actuates a log dial most conveniently placed in the Chartroom. This involves a permanent system of wiring to connect the chartroom indicators to the log position wherever it may be, and the coupling to the log mechanism is achieved by a socket and plug unit near the main log indicator.

Other Logs.—The patent log described above requires time to "stream" or to "haul in" and can become fouled, all of which require manual attention so that other types of log were developed to replace the line and rotator. The two main types are the pressure log using the principle of the pitot tube and a submerged impeller, both of which involve a tube being pushed out through the bottom of the vessel. The pushing out and the retracting to the stowed position is an easy operation which can be done either mechanically or by an electric motor and the electrical recording systems give not only the distance travelled, but also an instantaneous indication of the speed of the vessel through the water. Care has to be taken that the tubes are properly retracted before docking or when passing over a bar.

A more recent development giving considerable accuracy is the Walker's Manalog which operates by sensing a potential difference generated in water moving relative to a fixed magnetic field. The magnetic field is produced by an electro-magnet housed in the "sensing head" or probe which projects 22 inches (approximately 560 mm) through the bottom of the vessel whilst two electrodes flush with the surface of the probe, provide the means of measuring the potential difference which is proportional to the speed and the results are electronically recorded on a combined speed and distance repeater.

SOUNDING MACHINE.

The machine described is now obsolescent but it may still be found in some ships although no longer a requirement.

The deep-sea sounding machine consists of a small "hand winch" with a drum holding about 300 fathoms of piano wire from the end of which is secured a sinker or lead of about 28 lbs. weight. Handles are attached to the drum and can be used to release it so that the drum runs freely, or to activate a brake to stop the drum's rotation and by releasing a catch, the same handles can be used to wind the wire back on to the drum. Between the wire and the sinker is a two-fathom length of "stray line" of hemp to which is seized a perforated brass guard tube to accommodate and protect the sounding tube.

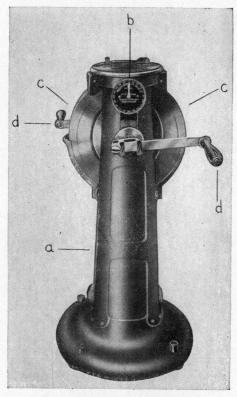

Fig. 4.—"Kelvite" Sounding Machine.

(a) Pedestal.
(b) Counter or Meter.
(c) Wire Drum.
(d) Winding and Brake Handles.

The machine is fitted amidships near the Navigating Bridge so that a boom is required to keep the wire and lead well clear of the ship's side. The boom is supported from a swivelling gooseneck by a topping lift, steadied by guys and has a carrier with a fairlead through which the sounding wire is rove. The carrier can be hauled out to the end of the boom by an outhaul and returned by means of an inhaul.

When the wire is running out, a metal "feeler" is held on it and a sudden slackening under the pressure of the hand indicates the sinker has reached the bottom. The brake is applied and the wire wound in to recover the sounding-tube from the brass guard-tube.

The principles involved are related to the fact that the pressure in water varies directly as the depth and to Boyle's Law which states that the volume of a given mass of gas is inversely proportional to the pressure providing the temperature is constant. The Kelvin chemical glass sounding-tube is 24 inches in length, of constant bore and sealed at one end whilst the inside of the tube is coated with chromate of silver which is red in colour. It is inserted into the guard-tube with the open down so that a column of air is trapped in the tube and as the volume decreases with increasing pressure, sea-water enters the tube as it descends causing the chromate to change to chloride of silver which is white in colour. The length of the discoloured part of the tube is a function of the depth to which it descended and the sounding in fathoms is obtained by placing the closed end of the tube up into a specially graduated boxwood scale and reading off the depth level with the line of demarcation. Should the boxwood scale be misplaced an approximate depth in fathoms is given by the following:—

$$\text{Depth in fathoms} = 5.5 \times \frac{\text{Discoloured length of tube}}{\text{Remaining length of tube}}$$

An advantage over the chemical tube is provided by the Wigzell tube in which a non-return valve traps the water as the tube descends and the sounding is again obtained by comparison with a boxwood scale, but by removing the water the same tube can be used over and over again.

Echo-sounding.—It may justly be said that the nearest land is usually underneath the ship's keel and it is therefore not surprising that one of the most important of the electronic navigational aids is the Echo-Sounder.

Most echo-sounding machines of recent design are so astonishingly reliable and simple to use, that the normal hand lead or deep-sea sounding-machine has been superseded. The range over which the echo-sounder can indicate depth is from two or three feet of water below the keel to 500 fathoms or more. An additional facility incorporated in most echo-sounding machines is that a continuous record of soundings may be obtained. These are shown as fine burn marks upon a continuous sheet of graph paper.

The principle upon which this apparatus works is that sound travels through water with a fairly uniform velocity of about 4,800 ft./sec. Sound pulses, sent out from the ship's bottom by means of an electrically operated transmitter, travel down to the sea bed and are reflected back towards the ship. The sound receiver, excited by these returning pulses, operates the stylus within the chartroom recorder. The distance travelled to the sea bed is directly related to the distance travelled across the graph paper by the stylus and marks are made on the recording paper accordingly.

METEOROLOGICAL EQUIPMENT.

The mariner in navigating his vessel between ports is loathe to take the ship unnecessarily into heavy weather. Although he is provided with weather reports from various sources by radio-communication, a local appraisal of the weather situation is frequently of much more value. The barometer, thermometer and hygrometer are the fundamental items of equipment used aboard ship for this purpose.

Barometer.—This instrument measures the pressure of the atmosphere at that point. Rapid changes in the pressure over a period of a few hours, or even less, are usually associated with wind. The mariner, by keeping an eye on the pressure changes at intervals throughout the day, is able to assess what sequence of weather he may expect, e.g. depression, cyclonic or anticyclonic conditions.

Hygrometer.—This consists of two identical thermometers hung side by side within a case, which screens them from the direct rays of the sun which would otherwise invalidate the readings obtained. The base of one thermometer is wrapped in muslin which is kept moist from water in a small jar beneath it. Whether it is a dry or damp day, the uncovered thermometer will indicate the correct air temperature. The thermometer wrapped in muslin will have its reading modified by the dryness of the surrounding air. On a dry day, the water on the muslin evaporates at a higher rate than on a moist damp day. Water evaporating from the muslin takes heat from the enclosed thermometer bulb, which is evident by the lowering in the indicated temperature. On a dry day, therefore, the difference in readings between the two thermometers will be greater than on a damp day. The immediate value of this knowledge to the navigator is to assess the likelihood of fog developing. A knowledge of the humidity (as this air dampness is called) is also of great value with regard to the ventilation of cargo holds.

Facsimile Reproduction Equipment.—This radio recorder apparatus is a recent innovation aboard ships although it has been in use ashore for many years. The present use of such equipment is to reproduce on board ship an exact replica of charted weather information propagated by Meteorological Offices at certain times of the day via local coast radio stations. Within the receiving apparatus aboard is a continuous roll of paper suitably marked by a moving stylus to correspond with information issued by the weather offices. The main advantage to the mariner is a sequence of charted weather information throughout the voyage which could otherwise only be obtained by the full-time employment of a specialist.

ELECTRONIC AND OTHER EQUIPMENT.

Radio Direction-finding.—(*See* Figure 5.) A mariner sighting a light from a lighthouse is able to say that his vessel is somewhere on a certain bearing from that particular object. Two lines of bearing from two different lighthouses would enable the mariner to "cross" the two bearings and so obtain his position. The radio

FIG. 5.—Marconi "Lodestar II" Automatic Direction-finder

direction-finding apparatus operates in much the same way but, since the waves approaching the vessel are invisible radio waves instead of visible light waves, a suitable means of receiving the waves is needed. This is achieved by a normal form of radio receiver. The aerial feeding this radio is of special construction so that is is possible to take bearings of particular radio transmitters sited ashore. The most modern types of D/F apparatus are able to take the bearings automatically if so desired. Direction-finding by means of radio waves is vulnerable to some forms of error and must be frequently checked.

V.H.F.—This consists of a radio receiver and transmitter aboard the vessel, receiving and transmitting the voice directly and thereby obviating the transmission of information by intermediaries such as the Ship's Radio Operator. By international convention a standard method of "calling up" stations has been agreed and apart from this the apparatus functions like an ordinary telephone. Since the apparatus operates on very high frequencies, the range of transmission is vaguely similar to the distance of the visible horizon.

Radar.—(*See* Figure 6.) This is probably the most important development in navigational aids in recent years. It paints on a screen a chart of all solid objects surrounding the ship. At the

FIG. 6.—"Decca 12 True Motion Radar Display" and "Automatic Relative Plotting Unit" (*left*).

user's choice, this "charted" range may be from a quarter mile up to 50 miles radius. Its greatest value may yet be as an aid to avoiding collision with other ships in the vicinity, although it is of proven value for position-finding when navigational marks or charted coastlines are within the range of the apparatus. Radio waves of a very short wavelength are sent out from the transmitting scanner in a bundle, which, upon reflection back to the ship after striking solid objects, cause temporary marks to be made on a cathode ray tube on the correct line of bearing and in relation to the range scale in use.

Decca Navigator.—An item of increasing navigational importance is the "Decca Navigator". This electronic aid enables a vessel to fix her position when other equipment is quite inadequate.

Fig. 7.—The Decca Navigator—Decometer Unit.

The simultaneous operation of a group of radio transmitters (four), located in known positions ashore, establishes a pattern, or lattice, of position lines in the atmosphere. Aboard the ship a special Decca receiving apparatus (Fig. 7) detects and indicates which lattice lines are intersecting at the vessel for the particular instant of noting the meter readings. These readings are intersected on ordinary navigational charts for that area, which have been overprinted with a red, green and purple hyperbolic lattice and the intersection of any two hyperbolic lines of different colour shows the ship's position.

Three of the four shore-based stations operate in unison with the fourth which is referred to as a Master Station whilst the former are usually referred to as Slave Stations. Each Slave Station pairs with the Master Station and is identified by the colour code of red, green and purple corresponding both with the identification meters on the Decca receiving apparatus and the three families of lattice lines overprinted on the chart.

Although the shore-based transmitters are very expensive to install and maintain, the apparatus aboard the ship is simple to use and can be relied upon to give an accurate position. This makes the system invaluable for coastal navigation and also when making a landfall.

FIG. 8.—Bridge of Modern Trawler.

1. Echo Sounder Recorders	4. Gyro Repeater
2. Decca Radar Display Unit	5. Decca Navigator
3. Decca Navigator Automatic Recorder	6. Steering Control Console

Chronometer.—A chronometer is an almost perfect timepiece which makes it possible to calculate the vessel's position from celestial observations. The mechanism is so finely balanced and compensated against temperature changes that the normal temperature variation or the operation of winding has little or no effect on the accuracy of its time-keeping. It is kept in a special compartment or case in a box and is mounted in gimbals to give full protection from shocks, vibrations or violent movement of any kind. When being carried ashore, its gimbals should be clamped.

So far as is possible, a chronometer should be wound by the same person every day at the same time and should be frequently checked by accurate time signals now available and results entered in a special book.

The introduction of the quartz crystal in place of the escapement wheel has led to the introduction of the electric chronometer, which only needs its battery changed once per year. It should still be checked against time signals daily.

Sextant.—The sextant is a hand-held precision instrument of double reflection roughly triangular in shape, which is used to measure the angular height of the sun or other celestial bodies above the visible horizon. During ocean passages the sextant, in conjunction with the chronometer and a current Nautical Almanac, gives a means of determining Position Lines from celestial observations and by combining two or more observations, the position of the ship can be determined in terms of latitude and longitude.

When a vessel is coasting, the sextant is useful for measuring angular heights of shore objects of known charted height to give the distance off and for taking horizontal angles between three charted objects to provide a fix.

Fig. 9.—Modern Bridge Layout.

Tachometer.—In all modern vessels this equipment is supplied as a bridge fitting in the wheelhouse or chartroom and is a replica of the one in the engine-room. It records the movement of the propeller shaft both in direction and the number of revolutions per minute. It gives an indication of the theoretical or approximate speed of the vessel through the water.

CHAPTER XII.

GENERAL SAFETY REGULATIONS

Rocket Life-saving Apparatus. Life-saving Equipment & Regulations. Fire-fighting. Lifebuoys and Lifebelts. Load-lines. Timber Deck Cargoes. Dock Regulations. Uniform System of Buoyage. Regulations for Preventing Collisions at Sea.

ROCKET LIFE-SAVING APPARATUS.

For the Guidance of Masters and Seamen when using the Rocket Apparatus for Saving Life.

Signals to be Used.

Signal	*Meaning*
1. Up and down movement of a white flag, white light or flare or the arms or a green star rocket.	Affirmative. Rocket line is held. Tail block made fast. Hawser made fast. Man in breeches buoy. Haul away.
2. Side to side movement of a white flag, white light or flare or of the arms fully extended or a red star rocket.	Negative. Slack away. Avast hauling.

Other signals used by life-saving organisations are given on page 175.

Procedure.

In the event of your vessel stranding on the coast of the United Kingdom and the lives of the crew being placed in danger, assistance will, if possible, be rendered from the shore in the following manner, namely:—

1. A rocket with a thin line attached will be fired across your vessel. Get hold of this line as soon as you can and then signal to the shore as indicated in (1) above.

   Alternatively, should your vessel use the Line Throwing Appliance to fire a line ashore, those on shore will secure a stouter line to it and make the signal (1) above. On seeing this signal haul in until the stouter line is on board and then signal as indicated in (1) above.

266

2. When the signal (1) is made from the shore, haul upon the rocket line until you get a tail block with an endless fall rove through it (the Whip).

3. Make fast the tail block close up to the mast or any convenient position which is accessible, bearing in mind that the lines must be clear from chafing and that space is left above the block for the hawser to be made fast. Unbend the rocket line from the whip and when this is all done, signal as in (1) above.

4. When the signal is seen on shore, a hawser will be bent to the whip line and will be hauled off to the ship by those on shore. A bowline will have been made with the end of the hawser round the hauling part of the whip except when there are rocks or other obstructions between the ship and the shore.

5. When the hawser is got on board, the bowline should be cast off and the end brought up between the two parts of the whip and made fast to the same part of the ship as the tail-block but just above.

6. When the hawser has been made fast, unbend the whip from the hawser, see that everything is clear and then signal to the shore as in (1) above.

7. The men on shore will then set the hawser taut and by means of the whip line will haul off to the ship the Breeches Buoy into which the person to be hauled ashore is to get. When he is in and secure, signal to the shore as in (1) above and the people on shore will then haul the person in the Breeches Buoy to the shore. When he is landed, the empty Breeches Buoy will be hauled back to the ship and the operation repeated until all are landed.

8. It may sometimes happen that the state of the weather and the condition of the ship will not admit of a hawser being set up and in such cases a Breeches Buoy will be hauled off by the whip which will be used without the hawser.

9. The system of signals must be strictly adhered to and it should be noted that the signal (1) above is made from the ship only when (a) the crew have got the rocket line, (b) when the tail-block has been made fast, (c) when the hawser has been made fast and (d) when a person is in the buoy ready to be hauled ashore. It is recommended that in order to facilitate the whole rescue operation, communication should be established whenever possible by semaphore or flashing lamp.

Masters and crews of stranded vessels should bear in mind that success in landing them depends in a great measure upon their own coolness and attention to the rules laid down.

FIRE-FIGHTING APPLIANCES.

(Rules in operation as from 25th May, 1980).

No matter what precautions are taken, there is always a risk of fire and it is the duty of every member of the crew of any vessel to be alive to this ever-present danger. He should know the drill of what to do in dealing with any outbreak however small which he may suddenly have to face and always see that everything possible is done to prevent such outbreaks.

In new construction the D.o.T. requirements ensure that set standards of fire-resisting materials are used in certain parts of the vessel, particularly in passageways to and from accommodation and working quarters.

The Statutory Instruments as published cover the requirements for all types and classes of ships but the following is given for the average vessel in the foreign-going trade excluding passenger ships where the Rules are much more stringent.

Boiler and Machinery Spaces.—Foreign-going vessels of 500 tons gross or over with oil-fired boilers or internal combustion machinery must have two fire hydrants, one to port and one to starboard, each with hose complete with coupling and dual purpose nozzle suitable for spraying water. Also a fixed fire-smothering gas or a pressurised water spray or a halogenated hydrocarbon vapourising liquid installation.

Ships using oil fuel only must have:—

(a) Receptacle containing ·3 cu. metre of sand and scoops.
(b) Two foam extinguishers of at least 10 gallons.
(c) Means for covering whole area of boiler-room to a depth of 6 inches with foam or fire-smothering gas in not more than five minutes.

Motor-ships must, in addition, have in the machinery spaces:—

(a) Foam fire extinguisher of at least 45 litres capacity.
(b) Portable foam extinguishers according to the distances involved.
(c) A foam applicator working from the fire main.

Deck Appliances.—Foreign-going vessels of 1,000 tons gross or over must carry:—

(a) Apparatus whereby smothering gas sufficient to give a minimum volume of free gas equal to 30 per cent of the gross volume of the largest hold in the ship, can be promptly conveyed by permanent piping system into any cargo compartment.

(b) Apparatus whereby at least two powerful jets of water can be rapidly and simultaneously directed into any part of the ship (under 1,000 tons gross only needs one jet). This includes two power pumps, a minimum of five hoses giving a total length of at least 60 per cent of the length of the vessel and an additional spare hose which must be available.

(c) Sufficient portable fire extinguishers (minimum of five).

(d) Two approved outfits (three if over 4,000 tons gross) consisting of breathing apparatus or smoke helmets, safety lamp and fireman's hatchet and also a portable electric drilling machine to provide emergency means of access to fires through deck or bulkhead.

Fire Buckets.—These are no longer required as part of the fire-fighting equipment except in the case of small vessels. Such buckets when carried are to be painted red and marked "Fire" and must be kept ready for use in accessible positions. Half must be fitted with lanyards.

Fire Hoses.—These shall be made of closely woven flax canvas or other suitable material and each shall be supplied with a plain nozzle (usually $\frac{1}{2}$ to $\frac{3}{4}$ in. in diameter) in addition to the 25 per cent which must have spray nozzles. They shall not exceed 60 feet in length unless the vessel's breadth is 90 feet or more when they may be 90 feet. They must be kept in a conspicuous position near the hydrant and must not be used for any other purpose. They are never rolled up like a deck hose but the coupling ends are placed together and the two parts rolled from the "bight". When required the hose can, by this method, be stretched very much more quickly.

Portable Fire Extinguishers.—These are normally of between 9 litres and 13·5 litres capacity and shall as far as practicable have a uniform method of operation.

If of the dry powder type they shall contain at least 4·5 kilogrammes of dry powder and shall not exceed in number half of the total extinguishers in any one space.

Spare charges must be provided for all types of portable extinguishers and they should be tested at intervals not exceeding four years. Full instructions for recharge are given with each type.

Portable Fire Extinguisher Colours.—It is suggested that in future the colour of the extinguisher shall indicate its contents as follows:—

Contents	Shell Colour
Water (Soda acid or CO_2)	Signal Red
Foam	Cream
Powder	French Blue
CO_2	Black
B.C.F. or B.T.M.	Emerald Green

The last two are liquids when under pressure as is CO_2, but they are in the group of Halogenated Hydrocarbous and suitable for use in radio or electrical fires.

Safety Lamp.—This must be of a type approved by the Mines Department of the D.T., and shall be operated by an electric battery. It must be capable of burning for a period of at least three hours.

Smoke Helmets and Breathing Apparatus.—These must be equipped with a fire-proof life and signalling line at least 10 feet longer than the air hose. It is attached by a snap hook to a strong leather or canvas safety belt or harness to be worn by the user of the helmet and the line shall be hemp-covered wire rope of $1\frac{1}{4}$ inches circumference (breaking strength $\frac{1}{2}$ ton). Plates of non-inflammable material with a legible code of signals clearly marked must be attached to the harness and to the free end of the lifeline.

If the apparatus is of the self-contained breathing apparatus type, it shall be of the open circuit compressed-air type with a storage capacity of at least 42 cu. ft. of free air. A pressure gauge shall be incorporated and means provided for audible warning when 80 per cent of the usable capacity has been consumed. The maximum weight shall not exceed 35 lbs. and a charged spare cylinder must be available together with a servicing and instructional manual.

LIFE-SAVING EQUIPMENT.

Lifebuoys.—An approved lifebuoy shall be capable of floating in fresh water for at least 24 hours with 32 pounds of iron suspended from it and must be constructed of solid cork or other equivalent. The inside diameter is 18 inches and the external diameter is 30 inches and they shall be painted a highly visible colour. The major axis of the section shall be 6 inches and the minor axis shall be 4 inches, whilst good beckets must be seized round the outside edge.

At least one on each side must be fitted with a life-line not less than 15 fathoms in length and a self-igniting light must be attached by 12 feet of good line to each of half the number of lifebuoys carried.

Both passenger and cargo vessels must carry a lifebuoy on each side of the navigating bridge in such a way that they can be promptly released to drop clear of the ship's side. Such lifebuoys must have attached to them self-igniting lights and self-activating smoke signals.

Foreign-going cargo vessels are required to carry not less than eight lifebuoys and in passengers' ships a larger number are carried depending upon the length of the vessel.

Using a Lifebuoy.—The easiest way to enter a lifebuoy in the water is to depress the nearest edge. The far side will then rise and drop over the head.

Lifejackets.—Every lifejacket for use by a person weighing 70 pounds or more shall provide a minimum of 35 pounds buoyancy in fresh water for 24 hours. They shall be marked indelibly on both sides in letters half-inch in size with the words "PERSON OF 70 LB. OR MORE" and on one side only the maker's name.

It shall be

(a) capable of being worn inside out;

(b) capable of turning the wearer to a safe floating position in still water within 5 seconds and support the head so that the mouth shall be not less than 6 inches above water;

(c) unaffected by oil or oil products;

(d) painted a highly visible colour;

(e) fitted with a strong ring or loop to facilitate rescue;

(f) fitted with an approved whistle firmly attached by a lanyard;

(g) supplied with fastening tapes to take a load of 200 pounds and shall allow the wearer to jump a vertical distance of 20 feet without injury.

The buoyancy shall be provided by kapok or its equivalent, and if of kapok it must contain not less than 35 ozs.

Lifejackets which depend upon inflation for buoyancy may be used by members of the crews of ships other than tankers, but if used they must have two separate buoyancy compartments and must be marked " CREW ONLY," and be capable of inflation both mechanically and by mouth.

Every lifejacket for use by a person weighing less than 70 pounds shall provide a minimum buoyancy of 15 pounds in fresh water for 24 hours. It must be indelibly marked "FOR PERSON UNDER 70 LB".

Life-lines.—During bad weather, life-lines are stretched along decks where there is any danger through water coming on board. To be effective they must not be stretched very tightly.

Line Throwing Appliances.—All vessels must be provided with an approved line-throwing appliance, capable of throwing a line $\frac{1}{2}$ inch in circumference, a distance of 250 yards in calm weather.

The apparatus must include four rockets and four lines each not less than 350 yards in length, with a breaking strain of not less than 250 lbs.

All such rocket gear must be kept in a watertight case.

Safety Net.—In many ports, regulations call for a rope net to be spread underneath the ship's gangway, and many lives have been saved by this measure.

Self-activating Smoke Signals.—These operate in a similar way to the Self-igniting Lights (see below) and must be capable of producing orange or other highly visible coloured smoke for a period of at least fifteen minutes.

Self-igniting Lights.—Small canisters which are attached to a lifebuoy by a length of line and are torn open by the act of throwing or releasing the lifebuoy overboard. The action of the water results in a flare which is such that it cannot be extinguished by the water, and they must burn for not less than 45 minutes.

The self-igniting lights attached to lifebuoys carried in tankers shall be of an electric battery type.

MERCHANT SHIPPING LOAD-LINE RULES 1968.

An International Load Line Certificate is required to be carried by practically all British ships and is issued to every ship which has been surveyed and marked in accordance with the International Convention on Load Lines, 1966. Though the Certificate remains in force for a period of up to 5 years, the vessel is liable for inspection at any time to see that the conditions are being complied with.

Vessels Exempted:—

      Ships of war.

      New ships of less than 24 metres (79 feet).

      Existing ships of less than 150 tons gross.

      Pleasure yachts not engaged in trade.

      Fishing vessels.

      Ships exempted by the D.O.T. for special reasons.

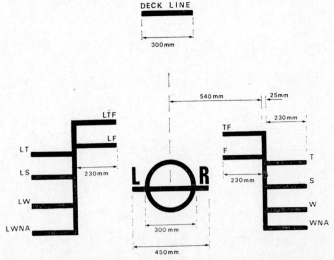

International Load Line Diagram (Starboard side).

International Load Line Certificate.—A certificate indicating the periodical approval by a classification society, authorised by the

B.O.T. to do so, is framed and hung in the chartroom or other conspicuous place.

Draught and Freeboard.—Before leaving any dock, wharf or harbour to proceed to sea, the master is required to enter draught and freeboard particulars in the log-book, and cause extracts to be exhibited in some place where they are accessible to the crew. Home Trade ships are not included in this Regulation.

Agreement with the Crew.—The master must insert particulars as to deck and load-lines in the "Agreement with the Crew", before any member of the crew signs the Agreement.

Markings.—Deck and load-lines are cut permanently into the ship's side, and are always painted either white or yellow on a dark background, or black on a light background. They are as follows:—

Deck Line.—This is the horizontal line 300 mm in length, marked on each side of the vessel to indicate the upper edge of the deck from which the freeboard is measured.

Load Line Mark.—This is still often referred to as "Plimsoll's Mark" and consists of a ring 300 mm in outside diameter, which is intersected by a horizontal line 450 mm in length, the upper edge of which passes through the centre of the ring which is fixed at a distance equal to the assigned summer freeboard measured vertically below the upper edge of the deck line. The freeboard is obtained from special Tables and takes into account the particular features of the ship and the trades for which she can be used.

Lines Used with the Load Line Mark.—540 mm forward of the centre of the ring is a vertical line to which a number of 230 mm horizontal lines are joined to indicate the greatest depth to which a vessel may be loaded under certain circumstances and in different seasons. These are as follows:—

Tropical Fresh Water Load Line	TF
Fresh Water Load Line ..	F
Tropical Load Line	T
Summer Load Line	S
Winter Load Line	W
Winter North Atlantic Load Line	WNA

NOTE.—(*a*) Sailing vessels only require in addition to the Load Line Mark, the Fresh Water Load Line and the Winter North Atlantic Load Line.

(*b*) If by reason of being a special vessel employed on some special service or where the navigational limits make any of the seasonal lines inappropriate, the lines may be omitted.

(*c*) All load line markings are 25 mm in breadth.

TIMBER DECK CARGOES.

Timber Load Lines.—There is an additional set of horizontal lines abaft the load line ring. This indicates the maximum load line for vessels loading timber in certain circumstances and seasons. The lettering is the same as for ordinary load lines, but the letter L is prefixed to each to indicate their special application.

Protection of Crew and Access to Machinery Space.—Deck cargoes must be stowed in such a manner that there is always a safe and satisfactory means of access to crews' quarters, machinery spaces, and other places used in the necessary work of the ship, and so that the said places may be properly closed and secured to keep out water.

A walkway not less than 1 metre in width shall be fitted over the deck cargo. It must be fitted with stanchions at least 1 metre high and not more than 1·5 metres apart. There must be three lines of guard rails or wires, the lowest must be not more than 230 mm from the level of the cargo and the other openings between the rails must be not more than 380 mm.

Lashings.—Throughout the length of the deck cargo, spaced not more than 3 metres (9·8 feet) apart there shall be independent overall lashings in good condition, consisting of close link chain not less than 19 millimetres ($\frac{3}{4}$ inch), or flexible wire rope of equal strength, and fitted with stretching screws and slip hooks which shall be accessible at all times. So that the length of the wire rope lashings may be regulated, they shall be fitted with a short length of long link chain.

The spacing of lashings shall be reduced to suit timber less than 3·6 metres (11·8 feet) long, or other suitable provision may be made.

Uprights.—If the nature of the timber is such that in order to ensure a secure stow, uprights are necessary, then they must fit into sockets of adequate strength and not more than three metres apart fore and aft.

SHIP MEASUREMENT.

The Ton Register.—The unit of capacity of a ship was originally the space occupied by a Last of 10 Quarters of Wheat, but is now taken as 100 English cubic feet. This is the standard ton register used by all maritime nations.

Cargo Measurement.—The space occupied by four hogsheads of wine was originally called a Tun and equalled 50 cubic feet. The ton measurement for cargo is now reckoned at 40 cubic feet, the approximate bulk of 4 quarters of a short ton of wheat.

Under Deck Tonnage.—The internal volume of a ship (100 cubic feet to the ton), between the under surface of the tonnage deck and the top of the ceiling or double bottom of the hold.

Gross Tonnage.—The total volume of a ship in cubic feet divided by 100. It includes most enclosed spaces above the tonnage deck.

Net Registered Tonnage.—From the total or gross tonnage of a vessel certain deductions are made on account of crew spaces, engine-room, water ballast and any space not used for passengers and cargo. The remainder is the net tonnage. *See* M.S. (Tonnage) Regulations 1982.

Deadweight Tonnage.—The actual carrying capacity in tons (2,240 lbs.) that a vessel is capable of carrying if loaded to her load water line with cargo, bunkers, stores and equipment.

Displacement Tonnage.—The weight of a vessel and contents in tons, or the number of tons of sea water displaced by a vessel when floating at her load water line.

DOCKS REGULATIONS.

These Regulations require every ship to carry a register in a form prescribed by the Home Office.

The register contains certificates of periodical tests and examinations of all machinery, permanent attachments, and gear connected with the loading or discharging of cargo. Any competent person may make these inspections, and the certificate must be signed by such person.

The Master or any other officer in charge of a vessel must produce the register to any of H.M. Inspectors of Factories when required.

Part 2 (9) Gangways.—A ship lying at a wharf or quay must have, when reasonably practicable, a properly secured gangway, not less than 22 inches wide, fenced on each side throughout with upper and lower rails, or taut ropes or chains, to a height of 2 feet 9 inches. In other cases, a ladder properly secured to prevent slipping, and of adequate length and sound material may be used.

Part 2 (10) Gangways for Vessels Lying Together.—When two vessels are lying alongside one another, and persons cannot pass freely from one to the other without undue risk the vessel with the higher freeboard shall provide the means of access.

Part 4 (35) Passages on Wharfs.—When goods are placed on a wharf or quay, a clear passage leading to the gangway or other means of access (required by Regulation 9) must be maintained, and when a space is left along the edge of a wharf or quay it must be at least 3 feet wide. Apart from fixed structures, plant and appliances in use, the passage shall be free of all obstructions.

Part 3 (8) Inspection of Machinery, Derricks and Bridle Chains.—Bridle chains attached to derricks, mast or deck, and all derricks or other permanent attachments shall be inspected once in every twelve months, and thoroughly examined every four years. Machinery shall be thoroughly examined at least once every twelve months.

Part 3 (19) Inspection of Chains, Rings, Hooks, Shackles and Swivels.—Chains, rings, hooks, shackles and swivels must not be used for hoisting or lowering until a certificate of test and examination has been obtained. All such gear, except bridle chains attached to derricks and masts, shall be examined by a competent person on each occasion before use, except when they have been inspected within the preceding three months. All gear used for hoisting or lowering which has been altered, repaired, or lengthened by welding, must be tested and re-examined.

Part 3 (20) Inspection of Ropes.—Ropes used for hoisting or lowering must be of suitable quality and free from patent defect, and for wire ropes used for the same purpose, a certificate giving certain details must be obtained from the manufacturers. All wire ropes in general use for the purpose of hoisting and lowering shall be inspected at least once every three months by a competent person, and once per month if any wire has broken in such rope. Such wire rope shall not be used if it shows signs of excessive wear, corrosion, or the number of visible broken wires in any length of eight diameters exceeds ten per cent of the total number of wires in the rope, and in the opinion of the person inspecting it is unfit for use. An eye splice made in any wire rope shall have three tucks with each whole strand of the rope, and two tucks with half of the wires cut out of each strand, and they must be tucked against the lay of the rope.

This Regulation does not prevent the use of any other type of splice which has proved as efficient as the one laid down in this Regulation.

Part 3 (23) Safe Working Load of Blocks.—Every gin, block or pulley used for hoisting or lowering, or other similar gear, shall have the safe working load clearly stamped upon it.

Part 3 (24) Marking of Chain and Wire Rope Slings.—So that any person may ascertain the safe working load of any chain or wire rope sling, the said particulars shall be marked in plain figures or letters upon the sling, or on a tablet or ring or durable material securely attached thereto, or by means of notices so exhibited as to be easily read by any person concerned.

Part 3 (25) Protection of Chain Slings.—Shortening slings by tying knots in them is prohibited. Where the links of a chain come in contact with sharp edges of a load and are liable to damage, packing must be provided.

Part 3 (26) Exposed Machinery.—All live electric conductors, motors, chain and friction gearing, shafting, or cog-wheels, shall (unless proved to be quite safe) be securely fenced without infringing any requirement of the Board of Trade, or impeding safe working of the ship.

Part 3 (27) Reversing Gear for Winches.—A suitable spring or other locking device must be provided on the lever controlling the link motion reversing gear of a winch.

Part 4 (34) Restriction on Age of Winch Drivers.—A person under sixteen years of age shall not be employed as a crane or winch driver, either on mechanical types or otherwise, nor must he give signals to any driver or attend falls on winch ends or bodies.

Part 4 (37) Fencing of Hatchways.—When the hatchways exceed 5 feet in depth, and are not in use for working cargo, coal or stores, or for the purpose of trimming, and the coamings are less than 2 feet 6 inches in height, they shall be fenced to a height of 3 feet or covered over with the proper hatch covers. This regulation does not apply during meal hours or other short periods while work is not in progress, or to trimming hatches which are not accessible to men working the cargo.

The use of hatch coverings for the construction of deck or cargo stages, or for any purpose which exposes them to damage is prohibited.

Hatch covers must be placed on the hatches in the positions indicated by the markings thereon.

Part 2 (11) Provision of Ladders and Cleats in Holds.—Where the depth of a hold exceeds 5 feet a ladder must be provided, or a means of access by cleats or cups on the coamings on bulkheads. Such ladders are not deemed to be safe unless an upper or lower section are in a vertical line, and not recessed under the deck more than is reasonably necessary to keep the hatchway clear. Cleats or cups on coamings, bulkheads or trunk hatchways must stand out not less than $4\frac{1}{2}$ inches for a width of 10 inches, and be so constructed as to prevent a man's foot slipping off the side.

Cargo must be stowed sufficiently far from the rungs of a ladder to allow room for a man's foot.

Part 4 (35) Working Cargo or Intermediate Decks.—If cargo is loaded or unloaded by means of a fall or sling at an intermediate deck, the hatch at that deck must be securely covered, or a landing platform placed across it with a width of not less than one section of hatch coverings. This regulation does not apply to any process of unloading which would be completed within a period of half an hour.

Part 4 (36) Restriction on Use of Hooks.—Except for the purpose of breaking out or making up slings, the practice of attaching hooks to the bands or fastenings of bales of cotton or similar goods is prohibited, when the working space of a hold is confined to the square of a hatch.

Part 4 (37) Working Cargo on Skeleton Deck.—When working on a skeleton deck staging must be provided, unless the space beneath is filled with cargo to within 2 feet of such deck.

Part 4 (38) Security of Hatch Beams.—If not removed, the beams of a hatch in use shall be adequately secured to prevent displacement.

Part 2 (13) Gear for Lifting Hatch Beams.—All fore and aft and athwartship beams, used for hatch covering, shall be provided with suitable gear for lifting on and off, to obviate the necessity of any person going upon them to adjust such gear.

Part 5 (43) Restriction Regarding Hatch Beams.—No person shall go upon, nor must any person order or authorise another person to go upon the fore and aft or athwartship beams, to adjust gear for the purpose of lifting them on or off.

Part 4 (39) Employment of Signallers at Hatchways.—If two (or more) falls are in use for the purpose of loading or unloading, a separate signaller shall be employed to attend to each fall.

MANNING OF SHIPS.

The Merchant Shipping Act provides for the detention of any vessel in any port of the United Kingdom as being unsafe if, by reason (amongst other things) of undermanning, the vessel is considered unfit to proceed to sea without serious danger to human life, having regard to the nature of the service for which the vessel is intended.

So far as the Deck Department is concerned, the number forming the deck complement (apart from the Master and Deck Officers) normally consists of a Boatswain together with a number of Able Seamen based on agreed official minimum scales depending upon the size and type of ship and a survey to show that the deck complement is adequate for all mooring and unmooring operations. This number varies from five in the smaller ships up to nine in the larger vessels but in all cases Efficient Deck Hands may be substituted for able seamen. Further substitutions may allow up to two Senior Ordinary Seamen in lieu of two of the Able Seamen but a third Able Seaman would require to be replaced by at least two Ordinary Seamen.

In the case of vessels fitted with an automatic pilot, steel hatches and an alarm system for summoning the watch below, a reduction of one or two Able Seamen is allowed (according to size of vessel) with the possible substitution of a Senior Ordinary Seaman for one of the Able Seamen and of two Ordinary Seamen for each of up to two more of the complement. Cadets may be included in the minimum requirements and in addition two Deck Boys may substitute for one Junior Ordinary Seaman or three for two.

The following definitions apply to these Manning Scales:—

(a) **Boatswain.**—A certificated able seaman of 20 years of age or over who can prove by discharges, indentures or other evidence that he has had at least four years' service at sea on deck.

(b) **Able Seaman.**—The holder of a certificate of competency as able seaman issued by, or recognised by the Department of Trade and Industry or a fourth year apprentice.

(c) **Efficient Deck Hand.**—The holder of a certificate of qualification as efficient deck hand.

(d) **Senior Ordinary Seaman.**—A seaman not less than $17\frac{1}{2}$ years of age who has had 18 months' sea service on deck, or a third year apprentice or, in ships over 5,500 gross tons, a carpenter provided he has had at least 12 months' sea service, or 24 months sea service as a G.P. rating.

(e) **Junior Ordinary Seaman.**—A seaman not less than $16\frac{1}{2}$ years of age who has had nine months' sea service on deck, or a second year apprentice or, a carpenter who does not fall within the definition in paragraph 9(d), or 12 months sea service as a G.P. rating.

(f) **Deck Boy.**—A boy who has served at sea on deck for less than nine months.

Present Trends in Manning.—In recent years the rapid development aboard ship of mechanization, automation and electronics have led to a reduction in the total size of the maritime labour force required for shipboard operation. This has meant the introduction of new manning systems agreed between the Shipping Company, the Unions and the D.o.T. on an experimental basis, allowing variations in the system from Company to Company but with agreed higher rates of pay. The two main systems are:—

(i) **Inter-Departmental Flexibility (I.D.F.)**—In this system all ratings are available to work for limited periods (in general not more than six hours per week) in other departments with duties arranged at meetings of the Ship's Management Committee. The work of the deck and engine-room ratings in departments other than their own, must be of a general or non-specialist nature so that no additional training or qualifications though encouraged are required to operate the system. Similarly with the catering department, except that the work is limited to general work on deck, such as cleaning and painting. All departments may be involved in duties associated with mooring and unmooring.

(ii) **General Purpose Rating.**—This system provides an integrated or non-departmental crew in that the deck, engine-room and catering ratings are combined into a general purpose labour force. It involves special training schemes which are at present mostly Company operated but it could lead to the development of approved courses suitable

K

from the beginning of training for all "seamen" who may form part of an integrated crew.

At present an ex-engine-room rating is required to attend at an approved engine-room/deck conversion course, after which he may take the qualifying examination for E.D.H. but the certificate would not be issued until the other conditions of steering experience and qualifying length of service as a general purpose rating have been completed. (See page 9.)

So far as the deck rating is concerned, special training for working in the engine-room is in the hands of the individual company and commences before joining any ship as part of an integrated crew. With this new approach to ship operation, there have been vast improvements in conditions and amenities, giving attractive career prospects, whilst some of the present nomenclatures are giving way to titles such as "Chief Petty Officer" (specialised trained for supervisory duties), "Petty Officer", "Leading Seaman", "Operator", etc. Some owners do not include the Catering department and hence the ratings of the deck and engine-room may form a joint labour force and become "Dual Purpose" ratings under somewhat similar conditions.

INTERNATIONAL REGULATIONS
FOR PREVENTING COLLISIONS AT SEA, 1972.
Including the 1981 Amendments.

PART A. GENERAL

RULE 1

Application

(*a*) These Rules shall apply to all vessels upon the high seas and in all waters connected therewith navigable by seagoing vessels.

(*b*) Nothing in these Rules shall interfere with the operation of special rules made by an appropriate authority for roadsteads, harbours, rivers, lakes or inland waterways connected with the high seas and navigable by seagoing vessels. Such special rules shall conform as closely as possible to these Rules.

(*c*) Nothing in these Rules shall interfere with the operation of any special rules made by the Government of any State with respect to additional station or signal lights, shapes or whistle signals for ships of war and vessels proceeding under convoy, or with respect to additional station or signal lights or shapes for fishing vessels engaged in fishing as a fleet. These additional station or signal lights, shapes or whistle signals shall, so far as possible, be such that they cannot be mistaken for any light, shape or signal authorized elsewhere under these Rules.

(*d*) Traffic separation schemes may be adopted by the Organization for the purpose of these Rules.

(*e*) Whenever the Government concerned shall have determined that a vessel of special construction or purpose cannot comply fully with the provisions of any of these Rules with respect to the number, position, range or arc of visibility of lights or shapes, as well as to the disposition and characteristics of sound-signalling appliances, without interfering with the special function of the vessel, such vessel shall comply with such other provisions in regard to the number, position, range or arc of visibility of lights or shapes, as well as to the disposition and characteristics of sound-signalling appliances, as her Government shall have determined to be the closest possible compliance with these Rules in respect to that vessel.

RULE 2

Responsibility

(*a*) Nothing in these Rules shall exonerate any vessel, or the owners, master or crew thereof, from the consequences of any neglect to comply with these Rules or of the neglect of any precaution which may be required by the ordinary practice of seamen, or by the special circumstances of the case.

(*b*) In construing and complying with these Rules due regard shall be had to all dangers of navigation and collision and to any special circumstances, including the limitations of the vessels involved, which may make a departure from these Rules necessary to avoid immediate danger.

RULE 3

General definitions

For the purpose of these Rules, except where the context otherwise requires:

(*a*) The word "vessel" includes every description of water craft, including non-displacement craft and seaplanes, used or capable of being used as a means of transportation on water.

(*b*) The term "power-driven vessel" means any vessel propelled by machinery.

(c) The term "sailing vessel" means any vessel under sail provided that propelling machinery, if fitted, is not being used.

(d) The term "vessel engaged in fishing" means any vessel fishing with nets, lines, trawls or other fishing apparatus which restrict manoeuvrability, but does not include a vessel fishing with trolling lines or other fishing apparatus which do not restrict manoeuvrability.

(e) The word "seaplane" includes any aircraft designed to manoeuvre on the water.

(f) The term "vessel not under command" means a vessel which through some exceptional circumstance is unable to manoeuvre as required by these Rules and is therefore unable to keep out of the way of another vessel.

(g) The term "vessel restricted in her ability to manoeuvre" means a vessel which from the nature of her work is restricted in her ability to manoeuvre as required by these Rules and is therefore unable to keep out of the way of another vessel.

The term "vessels restricted in their ability to manoeuvre" shall include but not be limited to:

   (i) a vessel engaged in laying, servicing or picking up a navigation mark, submarine cable or pipeline;

  (ii) a vessel engaged in dredging, surveying or underwater operations;

 (iii) a vessel engaged in replenishment or transferring persons, provisions or cargo while underway;

  (iv) a vessel engaged in the launching or recovery of aircraft;

   (v) a vessel engaged in mineclearance operations;

  (vi) a vessel engaged in a towing operation such as severely restricts the towing vessel and her tow in their ability to deviate from their course.

(h) The term "vessel constrained by her draught" means a power-driven vessel which because of her draught in relation to the available depth of water is severely restricted in her ability to deviate from the course she is following.

(i) The word "underway" means that a vessel is not at anchor, or made fast to the shore, or aground.

(j) The words "length" and "breadth" of a vessel mean her length overall and greatest breadth.

(k) Vessels shall be deemed to be in sight of one another only when one can be observed visually from the other.

(l) The term "restricted visibility" means any condition in which visibility is restricted by fog, mist, falling snow, heavy rainstorms, sandstorms or any other similar causes.

PART B. STEERING AND SAILING RULES

Section I. Conduct of vessels in any condition of visibility

RULE 4

Application

Rules in this Section apply in any condition of visibility.

RULE 5

Look-out

Every vessel shall at all times maintain a proper look-out by sight and hearing as well as by all available means appropriate in the prevailing circumstances and conditions so as to make a full appraisal of the situation and of the risk of collision.

Rule 6

Safe speed

Every vessel shall at all times proceed at a safe speed so that she can take proper and effective action to avoid collision and be stopped within a distance appropriate to the prevailing circumstances and conditions.

In determining a safe speed the following factors shall be among those taken into account:

(a) By all vessels:

 (i) the state of visibility;

 (ii) the traffic density including concentrations of fishing vessels or any other vessels;

 (iii) the manoeuvrability of the vessel with special reference to stopping distance and turning ability in the prevailing conditions;

 (iv) at night the presence of background light such as from shore lights or from back scatter of her own lights;

 (v) the state of wind, sea and current, and the proximity of navigational hazards;

 (vi) the draught in relation to the available depth of water.

(b) Additionally, by vessels with operational radar:

 (i) the characteristics, efficiency and limitations of the radar equipment;

 (ii) any constraints imposed by the radar range scale in use;

 (iii) the effect on radar detection of the sea state, weather and other sources of interference;

 (iv) the possibility that small vessels, ice and other floating objects may not be detected by radar at an adequate range;

 (v) the number, location and movement of vessels detected by radar;

 (vi) the more exact assessment of the visibility that may be possible when radar is used to determine the range of vessels or other objects in the vicinity.

Rule 7

Risk of collision

(a) Every vessel shall use all available means appropriate to the prevailing circumstances and conditions to determine if risk of collision exists. If there is any doubt such risk shall be deemed to exist.

(b) Proper use shall be made of radar equipment if fitted and operational, including long-range scanning to obtain early warning of risk of collision and radar plotting or equivalent systematic observation of detected objects.

(c) Assumptions shall not be made on the basis of scanty information, especially scanty radar information.

(d) In determining if risk of collision exists the following considerations shall be among those taken into account:

 (i) such risk shall be deemed to exist if the compass bearing of an approaching vessel does not appreciably change.

 (ii) such risk may sometimes exist even when an appreciable bearing change is evident, particularly when approaching a very large vessel or a tow or when approaching a vessel at close range.

K*

RULE 8

Action to avoid collision

(*a*) Any action taken to avoid collision shall, if the circumstances of the case admit, be positive, made in ample time and with due regard to the observance of good seamanship.

(*b*) Any alteration of course and/or speed to avoid collision shall, if the circumstances of the case admit, be large enough to be readily apparent to another vessel observing visually or by radar; a succession of small alterations of course and/or speed should be avoided.

(*c*) If there is sufficient sea room, alteration of course alone may be the most effective action to avoid a close-quarters situation provided that it is made in good time, is substantial and does not result in another close-quarters situation.

(*d*) Action taken to avoid collision with another vessel shall be such as to result in passing at a safe distance. The effectiveness of the action shall be carefully checked until the other vessel is finally past and clear.

(*e*) If necessary to avoid collision or allow more time to assess the situation, a vessel shall slacken her speed or take all way off by stopping or reversing her means of propulsion.

RULE 9

Narrow channels

(*a*) A vessel proceeding along the course of a narrow channel or fairway shall keep as near to the outer limit of the channel or fairway which lies on her starboard side as is safe and practicable.

(*b*) A vessel of less than 20 metres in length or a sailing vessel shall not impede the passage of a vessel which can safely navigate only within a narrow channel or fairway.

(*c*) A vessel engaged in fishing shall not impede the passage of any other vessel navigating within a narrow channel or fairway.

(*d*) A vessel shall not cross a narrow channel or fairway if such crossing impedes the passage of a vessel which can safely navigate only within such channel or fairway. The latter vessel may use the sound signal prescribed in Rule 34 (*d*) if in doubt as to the intention of the crossing vessel.

(*e*) (i) In a narrow channel or fairway when overtaking can take place only if the vessel to be overtaken has to take action to permit safe passing, the vessel intending to overtake shall indicate her intention by sounding the appropriate signal prescribed in Rule 34 (*c*) (i). The vessel to be overtaken shall, if in agreement, sound the appropriate signal prescribed in Rule 34 (*c*) (ii) and take steps to permit safe passing. If in doubt she may sound the signals prescribed in Rule 34 (*d*).

(ii) This Rule does not relieve the overtaking vessel of her obligation under Rule 13.

(*f*) A vessel nearing a bend or an area of a narrow channel or fairway where other vessels may be obscured by an intervening obstruction shall navigate with particular alertness and caution and shall sound the appropriate signal prescribed in Rule 34 (*e*).

(*g*) Any vessel shall, if the circumstances of the case admit, avoid anchoring in a narrow channel.

RULE 10

Traffic separation schemes

(*a*) This Rule applies to traffic separation schemes adopted by the Organization.

(*b*) A vessel using a traffic separation scheme shall:

    (i) proceed in the appropriate traffic lane in the general direction of traffic flow for that lane;

   (ii) so far as practicable keep clear of a traffic separation line or separation zone;

  (iii) normally join or leave a traffic lane at the termination of the lane, but when joining or leaving from either side shall do so at as small an angle to the general direction of traffic flow as practicable.

(*c*) A vessel shall so far as practicable avoid crossing traffic lanes, but if obliged to do so shall cross as nearly as practicable at right angles to the general direction of traffic flow.

(*d*) Inshore traffic zones shall not normally be used by through traffic which can safely use the appropriate traffic lane within the adjacent traffic separation scheme. However, vessels of less than 20 metres in length and sailing vessels may under all circumstances use inshore traffic zones.

(*e*) A vessel other than a crossing vessel or a vessel joining or leaving a lane shall not normally enter a separation zone or cross a separation line except:

    (i) in cases of emergency to avoid immediate danger;

   (ii) to engage in fishing within a separation zone.

(*f*) A vessel navigating in areas near the terminations of traffic separation schemes shall do so with particular caution.

(*g*) A vessel shall so far as practicable avoid anchoring in a traffic separation scheme or in areas near its terminations.

(*h*) A vessel not using a traffic separation scheme shall avoid it by as wide a margin as is practicable.

(*i*) A vessel engaged in fishing shall not impede the passage of any vessel following a traffic lane.

(*j*) A vessel of less than 20 metres in length or a sailing vessel shall not impede the safe passage of a power-driven vessel following a traffic lane.

(*k*) A vessel restricted in her ability to manoeuvre when engaged in an operation for the maintenance of safety of navigation in a traffic separation scheme is exempted from complying with this Rule to the extent necessary to carry out the operation.

(*l*) A vessel restricted in her ability to manoeuvre when engaged in an operation for the laying, servicing or picking up of a submarine cable, within a traffic separation scheme, is exempted from complying with this Rule to the extent necessary to carry out the operation.

Section II. Conduct of vessels in sight of one another

RULE 11

Application

Rules in this Section apply to vessels in sight of one another.

RULE 12

Sailing vessels

(*a*) When two sailing vessels are approaching one another, so as to involve risk of collision, one of them shall keep out of the way of the other as follows:

    (i) when each has the wind on a different side, the vessel which has the wind on the port side shall keep out of the way of the other;

   (ii) when both have the wind on the same side, the vessel which is to windward shall keep out of the way of the vessel which is to leeward;

  (iii) if a vessel with the wind on the port side sees a vessel to windward and cannot determine with certainty whether the other vessel has the wind on the port or on the starboard side, she shall keep out of the way of the other.

(*b*) For the purposes of this Rule the windward side shall be deemed to be the side opposite to that on which the mainsail is carried or, in the case of a square-rigged vessel, the side opposite to that on which the largest fore-and-aft sail is carried.

RULE 13

Overtaking

(*a*) Notwithstanding anything contained in the Rules of Part B, Sections I and II, any vessel overtaking any other shall keep out of the way of the vessel being overtaken.

(*b*) A vessel shall be deemed to be overtaking when coming up with another vessel from a direction more than 22·5 degrees abaft her beam, that is, in such a position with reference to the vessel she is overtaking, that at night she would be able to see only the sternlight of that vessel but neither of her sidelights.

(*c*) When a vessel is in doubt as to whether she is overtaking another, she shall assume that this is the case and act accordingly.

(*d*) Any subsequent alteration of the bearing between the two vessels shall not make the overtaking vessel a crossing vessel within the meaning of these Rules or relieve her of the duty of keeping clear of the overtaken vessel until she is finally past and clear.

RULE 14

Head-on situation

(*a*) When two power-driven vessels are meeting on reciprocal or nearly reciprocal courses so as to involve risk of collision each shall alter her course to starboard so that each shall pass on the port side of the other.

(*b*) Such a situation shall be deemed to exist when a vessel sees the other ahead or nearly ahead and by night she could see the masthead lights of the other in a line or nearly in a line and/or both sidelights and by day she observes the corresponding aspect of the other vessel.

(*c*) When a vessel is in any doubt as to whether such a situation exists she shall assume that it does exist and act accordingly.

RULE 15

Crossing situation

When two power-driven vessels are crossing so as to involve risk of collision, the vessel which has the other on her own starboard side shall keep out of the way and shall, if the circumstances of the case admit, avoid crossing ahead of the other vessel.

RULE 16

Action by give-way vessel

Every vessel which is directed to keep out of the way of another vessel shall, so far as possible, take early and substantial action to keep well clear.

RULE 17

Action by stand-on vessel

(*a*) (i) Where one of two vessels is to keep out of the way the other shall keep her course and speed.

(ii) The latter vessel may however take action to avoid collision by her manoeuvre alone, as soon as it becomes apparent to her that the vessel required to keep out of the way is not taking appropriate action in compliance with these Rules.

(*b*) When, from any cause, the vessel required to keep her course and speed finds herself so close that collision cannot be avoided by the action of the give-way vessel alone, she shall take such action as will best aid to avoid collision.

(*c*) A power-driven vessel which takes action in a crossing situation in accordance with sub-paragraph (*a*) (ii) of this Rule to avoid collision with another power-driven vessel shall, if the circumstances of the case admit, not alter course to port for a vessel on her own port side.

(*d*) This Rule does not relieve the give-way vessel of her obligation to keep out of the way.

Rule 18

Responsibilities between vessels

Except where Rules 9, 10 and 13 otherwise require:

(*a*) A power-driven vessel underway shall keep out of the way of:
  (i) a vessel not under command;
  (ii) a vessel restricted in her ability to manoeuvre;
  (iii) a vessel engaged in fishing;
  (iv) a sailing vessel.

(*b*) A sailing vessel underway shall keep out of the way of:
  (i) a vessel not under command;
  (ii) a vessel restricted in her ability to manoeuvre;
  (iii) a vessel engaged in fishing.

(*c*) A vessel engaged in fishing when underway shall, so far as possible, keep out of the way of:
  (i) a vessel not under command;
  (ii) a vessel restricted in her ability to manoeuvre.

(*d*) (i) Any vessel other than a vessel not under command or a vessel restricted in her ability to manoeuvre shall, if the circumstances of the case admit, avoid impeding the safe passage of a vessel constrained by her draught, exhibiting the signals in Rule 28.
  (ii) A vessel constrained by her draught shall navigate with particular caution having full regard to her special condition.

(*e*) A seaplane on the water shall, in general, keep well clear of all vessels and avoid impeding their navigation. In circumstances, however, where risk of collision exists, she shall comply with the Rules of this Part.

Section III. Conduct of vessels in restricted visibility

Rule 19

Conduct of vessels in restricted visibility

(*a*) This Rule applies to vessels not in sight of one another when navigating in or near an area of restricted visibility.

(*b*) Every vessel shall proceed at a safe speed adapted to the prevailing circumstances and conditions of restricted visibility. A power-driven vessel shall have her engines ready for immediate manoeuvre.

(*c*) Every vessel shall have due regard to the prevailing circumstances and conditions of restricted visibility when complying with the Rules of Section I of this Part.

(*d*) A vessel which detects by radar alone the presence of another vessel shall determine if a close-quarters situation is developing and/or risk of collision exists. If so, she shall take avoiding action in ample time, provided that when such action consists of an alteration of course, so far as possible the following shall be avoided:

(i) an alteration of course to port for a vessel forward of the beam, other than for a vessel being overtaken;

(ii) an alteration of course towards a vessel abeam or abaft the beam.

(*e*) Except where it has been determined that a risk of collision does not exist, every vessel which hears apparently forward of her beam the fog signal of another vessel, or which cannot avoid a close-quarters situation with another vessel forward of her beam, shall reduce her speed to the minimum at which she can be kept on her course. She shall if necessary take all her way off and in any event navigate with extreme caution until danger of collision is over.

PART C. LIGHTS AND SHAPES

RULE 20

Application

(*a*) Rules in this Part shall be complied with in all weathers.

(*b*) The Rules concerning lights shall be complied with from sunset to sunrise, and during such times no other lights shall be exhibited, except such lights as cannot be mistaken for the lights specified in these Rules or do not impair their visibility or distinctive character, or interfere with the keeping of a proper look-out.

(*c*) The lights prescribed by these Rules shall, if carried, also be exhibited from sunrise to sunset in restricted visibility and may be exhibited in all other circumstances when it is deemed necessary.

(*d*) The Rules concerning shapes shall be complied with by day.

(*e*) The lights and shapes specified in these Rules shall comply with the provisions of Annex I to these Regulations.

RULE 21

Definitions

(*a*) "Masthead light" means a white light placed over the fore and aft centre-line of the vessel showing an unbroken light over an arc of the horizon of 225 degrees and so fixed as to show the light from right ahead to 22·5 degrees abaft the beam on either side of the vessel.

(*b*) "Sidelights" means a green light on the starboard side and a red light on the port side each showing an unbroken light over an arc of the horizon of 112·5 degrees and so fixed as to show the light from right ahead to 22·5 degrees abaft the beam on its respective side. In a vessel of less than 20 metres in length the sidelights may be combined in one lantern carried on the fore and aft centre-line of the vessel.

(*c*) "Sternlight" means a white light placed as nearly as practicable at the stern showing an unbroken light over an arc of the horizon of 135 degrees and so fixed as to show the light 67·5 degrees from right aft on each side of the vessel.

(*d*) "Towing light" means a yellow light having the same characteristics as the "sternlight" defined in paragraph (*c*) of this Rule.

(*e*) "All round light" means a light showing an unbroken light over an arc of the horizon of 360 degrees.

(*f*) "Flashing light" means a light flashing at regular intervals at a frequency of 120 flashes or more per minute.

RULE 22

Visibility of lights

The lights prescribed in these Rules shall have an intensity as specified in Section 8 of Annex I to these Regulations so as to be visible at the following minimum ranges:

(*a*) In vessels of 50 metres or more in length:
—a masthead light, 6 miles;
—a sidelight, 3 miles;
—a sternlight, 3 miles;
—a towing light, 3 miles;
—a white, red, green or yellow all-round light, 3 miles.

(*b*) In vessels of 12 metres or more in length but less than 50 metres in length:
—a masthead light, 5 miles; except that where the length of the vessel is less than 20 metres, 3 miles;
—a sidelight, 2 miles;
—a sternlight, 2 miles;
—a towing light, 2 miles;
—a white, red, green or yellow all-round light, 2 miles.

(*c*) In vessels of less than 12 metres in length:
—a masthead light, 2 miles;
—a sidelight, 1 mile;
—a sternlight, 2 miles;
—a towing light, 2 miles;
—a white, red, green or yellow all-round light, 2 miles.

(*d*) In inconspicuous, partly submerged vessels or objects being towed:
—a white all-round light, 3 miles.

RULE 23
Power-driven vessels underway

(*a*) A power-driven vessel underway shall exhibit:
 (i) a masthead light forward;
 (ii) a second masthead light abaft of and higher than the forward one; except that a vessel of less than 50 metres in length shall not be obliged to exhibit such light but may do so;
 (iii) sidelights;
 (iv) a sternlight.

(*b*) An air-cushion vessel when operating in the non-displacement mode shall, in addition to the lights prescribed in paragraph (*a*) of this Rule, exhibit an all-round flashing yellow light.

(*c*) (i) A power-driven vessel of less than 12 metres in length may in lieu of the lights prescribed in paragraph (*a*) of this Rule exhibit an all-round white light and sidelights;
  (ii) a power-driven vessel of less than 7 metres in length whose maximum speed does not exceed 7 knots may in lieu of the lights prescribed in paragraph (*a*) of this Rule exhibit an all-round white light and shall, if practicable, also exhibit sidelights;
  (iii) the masthead light or all-round white light on a power-driven vessel of less than 12 metres in length may be displaced from the fore and aft centreline of the vessel if centreline fitting is not practicable, provided that the sidelights are combined in one lantern which shall be carried on the fore and aft centreline of the vessel or located as nearly as practicable in the same fore and aft line as the masthead light or the all-round white light.

RULE 24
Towing and pushing

(*a*) A power-driven vessel when towing shall exhibit:
 (i) instead of the light prescribed in Rule 12 (*a*) (i) or (*a*) (ii), two masthead lights in a vertical line. When the length of the tow, measuring from the stern of the towing vessel to the after end of the tow exceeds 200 metres, three such lights in a vertical line

(ii) sidelights;

(iii) a sternlight;

(iv) a towing light in a vertical line above the sternlight;

(v) when the length of the tow exceeds 200 metres, a diamond shape where it can best be seen.

(*b*) When a pushing vessel and a vessel being pushed ahead are rigidly connected in a composite unit they shall be regarded as a power-driven vessel and exhibit the lights prescribed in Rule 23.

(*c*) A poer-driven vessel when pushing ahead or towing alongside, except in the case of a composite unit, shall exhibit:

(i) instead of the light prescribed in Rule 23 (*a*) (i), two masthead lights forward in a vertical line;

(ii) sidelights;

(iii) a sternlight.

(*d*) A power-driven vessel to which paragraph (*a*) or (*c*) of this Rule applies shall also comply with Rule 23 (*a*) (ii).

(*e*) A vessel or object being towed, other than those mentioned in paragraph (*g*) of this Rule, shall exhibit:

(i) sidelights;

(ii) a sternlight;

(iii) when the length of the tow exceeds 200 metres, a diamond shape where it can best be seen.

(*f*) Provided that any number of vessels being towed alongside or pushed in a group shall be lighted as one vessel,

(i) a vessel being pushed ahead, not being part of a composite unit, shall exhibit at the forward end, sidelights;

(ii) a vessel being towed alongside shall exhibit a sternlight and at the forward end, sidelights.

(*g*) An inconspicuous, partly submerged vessel or object, or combination of such vessels or objects being towed, shall exhibit:

(i) if it is less than 25 metres in breadth, one all-round white light at or near the forward end and one at or near the after end except that dracones need not exhibit a light at or near the forward end;

(ii) if it is 25 metres or more in breadth, two additional all-round white lights at or near the extremities of its breadth;

(iii) if it exceeds 100 metres in length, additional all-round white lights between the lights prescribed in sub-paragraphs (i) and (ii) so that the distance between the lights shall not exceed 100 metres;

(iv) a diamond shape at or near the aftermost extremity of the last vessel or object being towed and if the length of the tow exceeds 200 metres an additional diamond shape where it can best be seen and located as far forward as is practicable.

(*h*) Where from any sufficient cause it is impracticable for a vessel or object being towed to exhibit the lights or shapes prescribed in paragraph (*e*) or (*g*) of this Rule, all possible measures shall be taken to light the vessel or object towed or at least to indicate the presence of such vessel or object.

(*i*) Where from any sufficient cause it is impracticable for a vessel not normally engaged in towing operations to display the lights prescribed in paragraph (*a*) or (*c*) of this Rule, such vessel shall not be required to exhibit those lights when engaged in towing another vessel in distress or otherwise in need of assistance. All possible measures shall be taken to indicate the nature of the relationship between the towing vessel and the vessel being towed as authorized by Rule 36, in particular by illuminating the towline.

RULE 25

Sailing vessels underway and vessels under oars

(*a*) A sailing vessel underway shall exhibit:

(i) sidelights;

(ii) a sternlight.

(*b*) In a sailing vessel of less than 20 metres in length the lights prescribed in paragraph (*a*) of this Rule may be combined in one lantern carried at or near the top of the mast where it can best be seen.

(*c*) A sailing vessel underway may, in addition to the lights prescribed in paragraph (*a*) of this Rule, exhibit at or near the top of the mast, where they can best be seen, two all-round lights in a vertical line, the upper being red and the lower green, but these lights shall not be exhibited in conjunction with the combined lantern permitted by paragraph (*b*) of this Rule.

(*d*) (i) A sailing vessel of less than 7 metres in length shall, if practicable, exhibit the lights prescribed in paragraphs (*a*) or (*b*) of this Rule, but if she does not, she shall have ready at hand an electric torch or lighted lantern showing a white light which shall be exhibited in sufficient time to prevent collision.

(ii) A vessel under oars may exhibit the lights prescribed in this Rule for sailing vessels, but if she does not, she shall have ready at hand an electric torch or lighted lantern showing a white light which shall be exhibited in sufficient time to prevent collision.

(*e*) A vessel proceeding under sail when also being propelled by machinery shall exhibit forward where it can best be seen a conical shape, apex downwards.

Rule 26

Fishing vessels

(*a*) A vessel engaged in fishing, whether underway or at anchor, shall exhibit only the lights and shapes prescribed in this Rule.

(*b*) A vessel when engaged in trawling, by which is meant the dragging through the water of a dredge net or other apparatus used as a fishing appliance, shall exhibit:

(i) two all-round lights in a vertical line, the upper being green and the lower white, or a shape consisting of two cones with their apexes together in a vertical line one above the other; a vessel of less than 20 metres in length may instead of this shape exhibit a basket;

(ii) a masthead light abaft of and higher than the all-round green light; a vessel of less than 50 metres in length shall not be obliged to exhibit such a light but may do so;

(iii) when making way through the water, in addition to the lights prescribed in this paragraph, sidelights and a sternlight.

(*c*) A vessel engaged in fishing, other than trawling, shall exhibit:

(i) two all-round lights in a vertical line, the upper being red and the lower white, or a shape consisting of two cones with apexes together in a vertical line one above the other; a vessel of less than 20 metres in length may instead of this shape exhibit a basket;

(ii) when there is outlying gear extending more than 150 metres horizontally from the vessel, an all-round white light or a cone apex upwards in the direction of the gear;

(iii) when making way through the water, in addition to the lights prescribed in this paragraph, sidelights and a sternlight.

(*d*) A vessel engaged in fishing in close proximity to other vessels engaged in fishing may exhibit the additional signals described in Annex II to these Regulations.

(*e*) A vessel when not engaged in fishing shall not exhibit the lights or shapes prescribed in this Rule, but only those prescribed for a vessel of her length.

Rule 27

Vessels not under command or restricted in their ability to manoeuvre

(*a*) A vessel not under command shall exhibit:

  (i) two all-round red lights in a vertical line where they can best be seen;

  (ii) two balls or similar shapes in a vertical line where they can best be seen;

  (iii) when making way through the water, in addition to the lights prescribed in this paragraph, sidelights and a sternlight.

(*b*) A vessel restricted in her ability to manoeuvre, except a vessel engaged in mineclearance operations, shall exhibit:

  (i) three all-round lights in a vertical line where they can best be seen. The highest and lowest of these lights shall be red and the middle light shall be white;

  (ii) three shapes in a vertical line where they can best be seen. The highest and lowest of these shapes shall be balls and the middle one a diamond;

  (iii) when making way through the water, a masthead light or lights, sidelights and a sternlight, in addition to the lights prescribed in subparagraph (i);

  (iv) when at anchor, in addition to the lights or shapes prescribed in subparagraphs (i) and (ii), the light, lights or shape prescribed in Rule 30.

(*c*) A power-driven vessel engaged in a towing operation such as severely restricts the towing vessel and her tow in their ability to deviate from their course shall, in addition to the lights or shapes prescribed in Rule 24 (*a*), exhibit the lights or shapes prescribed in sub-paragraphs (*b*) (i) and (ii) of this Rule.

(*d*) A vessel engaged in dredging or underwater operations, when restricted in her ability to manoeuvre, shall exhibit the lights and shapes prescribed in sub-paragraphs (*b*) (i), (ii) and (iii) of this Rule and shall in addition, when obstruction exists, exhibit:

  (i) two all-round red lights or two balls in a vertical line to indicate the side on which the obstruction exists;

  (ii) two all-round green lights or two diamonds in a vertical line to indicate the side on which another vessel may pass;

  (iii) when at anchor, the lights or shapes prescribed in this paragraph instead of the lights or shape prescribed in Rule 30.

(*e*) Whenever the size of a vessel engaged in diving operations makes it impracticable to exhibit all lights and shapes prescribed in paragraph (*d*) of this Rule, the following shall be exhibited:

  (i) three all-round lights in a vertical line where they can best be seen. The highest and lowest of these lights shall be red and the middle light shall be white;

  (ii) a rigid replica of the International Code flag "A" not less than 1 metre in height. Measures shall be taken to ensure its all-round visibility.

(*f*) A vessel engaged in mineclearance operations shall in addition to the lights prescribed for a power-driven vessel in Rule 23 or to the lights or shape prescribed for a vessel at anchor in Rule 30 as appropriate, exhibit three all-round green lights or three balls. One of these lights or shapes shall be exhibited near the foremast head and one at each end of the fore yard. These lights or shapes indicate that it is dangerous for another vessel to approach within 1000 metres of the mineclearance vessel.

(*g*) Vessels of less than 12 metres in length, except those engaged in diving operations, shall not be required to exhibit the lights and shapes prescribed in this Rule.

(*h*) The signals prescribed in this Rule are not signals of vessels in distress and requiring assistance. Such signals are contained in Annex IV to these Regulations.

Rule 28
Vessels constrained by their draught

A vessel constrained by her draught may, in addition to the lights prescribed for power-driven vessels in Rule 23, exhibit where they can best be seen three all-round red lights in a vertical line, or a cylinder.

Rule 29
Pilot vessels

(*a*) A vessel engaged on pilotage duty shall exhibit:
  (i) at or near the masthead, two all-round lights in a vertical line, the upper being white and the lower red;
  (ii) when underway, in addition, sidelights and a sternlight;
  (iii) when at anchor, in addition to the lights prescribed in sub-paragraph (i), the light, lights or shape prescribed in Rule 30 for vessels at anchor.

(*b*) A pilot vessel when not engaged on pilotage duty shall exhibit the lights or shapes prescribed for a similar vessel of her length.

Rule 30
Anchored vessels and vessels aground

(*a*) A vessel at anchor shall exhibit where it can best be seen:
  (i) in the fore part, an all-round white light or one ball;
  (ii) at or near the stern and at a lower level than the light prescribed in sub-paragraph (i), an all-round white light.

(*b*) A vessel of less than 50 metres in length may exhibit an all-round white light where it can best be seen instead of the lights prescribed in paragraph (*a*) of this Rule.

(*c*) A vessel at anchor may, and a vessel of 100 metres and more in length shall, also use the available working or equivalent lights to illuminate her decks.

(*d*) A vessel aground shall exhibit the lights prescribed in paragraphs (*a*) or (*b*) of this Rule and in addition, where they can best be seen:
  (i) two all-round red lights in a vertical line;
  (ii) three balls in a vertical line.

(*e*) A vessel of less than 7 metres in length, when at anchor, not in or near a narrow channel, fairway or anchorage, or where other vessels normally navigate, shall not be required to exhibit the lights or shapes prescribed in paragraphs (*a*) and (*b*) of this Rule.

(*f*) A vessel of less than 12 metres in length, when aground, shall not be required to exhibit the lights or shapes prescribed in sub-paragraphs (*d*) (i) and (ii) of this Rule.

Rule 31
Seaplanes

Where it is impracticable for a seaplane to exhibit lights and shapes of the characteristics or in the positions prescribed in the Rules of this Part she shall exhibit lights and shapes as closely similar in characteristics and position as is possible.

PART D. SOUND AND LIGHT SIGNALS

Rule 32
Definitions

(*a*) The word "whistle" means any sound signalling appliance capable of producing the prescribed blasts and which complies with the specifications in Annex III to these Regulations.

(*b*) The term "short blast" means a blast of about one second's duration.

(*c*) The term "prolonged blast" means a blast of from four to six seconds' duration.

RULE 33

Equipment for sound signals

(*a*) A vessel of 12 metres or more in length shall be provided with a whistle and a bell and a vessel of 100 metres or more in length shall, in addition, be provided with a gong, the tone and sound of which cannot be confused with that of the bell. The whistle, bell and gong shall comply with the specifications in Annex III to these Regulations. The bell or gong or both may be replaced by other equipment having the same respective sound characteristics, provided that manual sounding of the prescribed signals shall always be possible.

(*b*) A vessel of less than 12 metres in length shall not be obliged to carry the sound signalling appliances prescribed in paragraph (*a*) of this Rule but if she does not, she shall be provided with ome other means of making an efficient sound signal.

RULE 34

Manoeuvring and warning signals

(*a*) When vessels are in sight of one another, a power-driven vessel underway, when manoeuvring as authorized or required by these Rules, shall indicate that manoeuvre by the following signals on her whistle:

—one short blast to mean "I am altering my course to starboard";

—two short blasts to mean "I am altering my course to port";

—three short blasts to mean "I am operating astern propulsion".

(*b*) Any vessel may supplement the whistle signals prescribed in paragraph (*a*) of this Rule by light signals, repeated as appropriate, whilst the manoeuvre is being carried out:

(i) these light signals shall have the following significance:

—one flash to mean "I am altering my course to starboard";

—two flashes to mean "I am altering my course to port";

—three flashes to mean "I am operating astern propulsion";

(ii) the duration of each flash shall be about one second, the interval between flashes shall be about one second, and the interval between successive signals shall be not less than ten seconds;

(iii) the light used for this signal shall, if fitted, be an all-round white light, visible at a minimum range of 5 miles, and shall comply with the provisions of Annex I to these Regulations.

(*c*) When in sight of one another in a narrow channel or fairway:

(i) a vessel intending to overtake another shall in compliance with Rule 9 (*e*) (i) indicate her intention by the following signals on her whistle:

—two prolonged blasts followed by one short blast to mean "I intend to overtake you on your starboard side";

—two prolonged blasts followed by two short blasts to mean "I intend to overtake you on your port side".

(ii) the vessel about to be overtaken when acting in accordance with Rule 9 (*e*) (i) shall indicate her agreement by the following signal on her whistle:

—one prolonged, one short, one prolonged and one short blast, in that order.

(*d*) When vessels in sight of one another are approaching each other and from any cause either vessel fails to understand the intentions or actions of the other, or is in doubt whether sufficient action is being taken by the other to avoid collision, the vessel in doubt shall immediately indicate such doubt by giving at least five short and rapid blasts on the whistle. Such signal may be supplemented by a light signal of at least five short and rapid flashes.

(*e*) A vessel nearing a bend or an area of a channel or fairway where other vessels may be obscured by an intervening obstruction shall sound one prolonged blast. Such signal shall be answered with a prolonged blast by any approaching vessel that may be within hearing around the bend or behind the intervening obstruction.

(*f*) If whistles are fitted on a vessel at a distance apart of more than 100 metres, one whistle only shall be used for giving manoeuvring and warning signals.

RULE 35
Sound signals in restricted visibility

In or near an area of restricted visibility, whether by day or night, the signals prescribed in this Rule shall be used as follows:

(*a*) A power-driven vessel making way through the water shall sound at intervals of not more than 2 minutes one prolonged blast.

(*b*) A power-driven vessel underway but stopped and making no way through the water shall sound at intervals of not more than 2 minutes two prolonged blasts in succession with an interval of about 2 seconds between them.

(*c*) A vessel not under command, a vessel restricted in her ability to manoeuvre, a vessel constrained by her draught, a sailing vessel, a vessel engaged in fishing and a vessel engaged in towing or pushing another vessel shall, instead of the signals prescribed in paragraphs (*a*) or (*b*) of this Rule, sound at intervals of not more than 2 minutes three blasts in succession, namely one prolonged followed by two short blasts.

(*d*) A vessel engaged in fishing, when at anchor, and a vessel restricted in her ability to manoeuvre when carrying out her work at anchor, shall instead of the signals prescribed in paragraph (*g*) of this Rule sound the signal prescribed in paragraph (*c*) of this Rule.

(*e*) A vessel towed or if more than one vessel is towed the last vessel of the tow, if manned, shall at intervals of not more than 2 minutes sound four blasts in succession, namely one prolonged followed by three short blasts. When practicable, this signal shall be made immediately after the signal made by the towing vessel.

(*f*) When a pushing vessel and a vessel being pushed ahead are rigidly connected in a composite unit they shall be regarded as a power-driven vessel and shall give the signals prescribed in paragraphs (*a*) or (*b*) of this Rule.

(*g*) A vessel at anchor shall at intervals of not more than one minute ring the bell rapidly for about 5 seconds. In a vessel of 100 metres or more in length the bell shall be sounded in the forepart of the vessel and immediately after the ringing of the bell the gong shall be sounded rapidly for about 5 seconds in the after part of the vessel. A vessel at anchor may in addition sound three blasts in succession, namely one short, one prolonged and one short blast, to give warning of her position and of the possibility of collision to an approaching vessel.

(*h*) A vessel aground shall give the bell signal and if required the gong signal prescribed in paragraph (*f*) of this Rule and shall, in addition, give three separate and distinct strokes on the bell immediately before and after the rapid ringing of the bell. A vessel aground may in addition sound an appropriate whistle signal.

(*i*) A vessel of less than 12 metres in length shall not be obliged to give the above-mentioned signals but, if she does not, shall make some other efficient sound signal at intervals of not more than 2 minutes.

(*j*) A pilot vessel when engaged on pilotage duty may in addition to the signals prescribed in paragraphs (*a*), (*b*) or (*f*) of this Rule sound an identity signal consisting of four short blasts.

RULE 36

Signals to attract attention

If necessary to attract the attention of another vessel any vessel may make light or sound signals that cannot be mistaken for any signal authorized elsewhere in these Rules, or may direct the beam of her searchlight in the direction of the danger, in such a way as not to embarrass any vessel.

Any light to attract the attention of another vessel shall be such that it cannot be mistaken for any aid to navigation. For the purpose of this Rule the use of high intensity intermittent or revolving lights, such as strobe lights, shall be avoided.

RULE 37

Distress signals

When a vessel is in distress and requires assistance she shall use or exhibit the signals described in Annex IV to these Regulations.

PART E.   EXEMPTIONS

RULE 38

Exemptions

Any vessel (or class of vessels) provided that she complies with the requirements of the International Regulations for Preventing Collisions at Sea, 1960, the keel of which is laid or which is at a corresponding stage of construction before the entry into force of these Regulations may be exempted from compliance therewith as follows:

(*a*) The installation of lights with ranges prescribed in Rule 22, until four years after the date of entry into force of these Regulations.

(*b*) The installation of lights with colour specifications as prescribed in Section 7 of Annex I to these Regulations, until four years after the date of entry into force of these Regulations.

(*c*) The repositioning of lights as a result of conversion from Imperial to metric units and rounding off measurement figures, permanent exemption.

(*d*) (i) The repositioning of masthead lights on vessels of less than 150 metres in length, resulting from the prescriptions of Section 3 (*a*) of Annex I to these Regulations, permanent exemption.

  (ii) The repositioning of masthead lights on vessels of 150 metres or more in length, resulting from the prescriptions of Section 3 (*a*) of Annex I to these Regulations, until nine years after the date of entry into force of these Regulations.

(*e*) The repositioning of masthead lights resulting from the prescriptions of Section 2 (*b*) of Annex I, until nine years after the date of entry into force of these Regulations.

(*f*) The repositioning of sidelights resulting from the prescriptions of Section 2 (*g*) and 3 (*b*) of Annex I, until nine years after the date of entry into force of these Regulations.

(*g*) The requirements for sound signal appliances prescribed in Annex III until nine years after the date of entry into force of these Regulations.

(*h*) The repositioning of all-round lights resulting from the prescription of Section 9 (*b*) of Annex I to these Regulations, permanent exemption.

ANNEX I

Positioning and technical details of lights and shapes

1. *Definition*

The term "height above the hull" means height above the uppermost continuous deck. This height shall be measured from the position vertically beneath the location of the light.

2. *Vertical positioning and spacing of lights*

(*a*) On a power-driven vessel of 20 metres or more in length the masthead lights shall be placed as follows:
  (i) the forward masthead light, or if only one masthead light is carried, then that light, at a height above the hull of not less than 6 metres, and, if the breadth of the vessel exceeds 6 metres, then at a height above the hull not less than such breadth, so however that the light need not be placed at a greater height above the hull than 12 metres;
  (ii) when two masthead lights are carried the after one shall be at least 4·5 metres vertically higher than the forward one.

(*b*) The vertical separation of masthead lights of power-driven vessels shall be such that in all normal conditions of trim the after light will be seen over and separate from the forward light at a distance of 1,000 metres from the stem when viewed from sea level.

(*c*) The masthead light of a power-driven vessel of 12 metres but less than 20 metres in length shall be placed at a height above the gunwale of not less than 2·5 metres.

(*d*) A power-driven vessel of less than 12 metres in length may carry the uppermost light at a height of less than 2·5 metres above the gunwale. When however a masthead light is carried in addition to sidelights and a sternlight, then such masthead light shall be carried at least 1 metre higher than the sidelights.

(*e*) One of the two or three masthead lights prescribed for a power-driven vessel when engaged in towing or pushing another vessel shall be placed in the same position as either the forward masthead light or the after masthead light; provided that, if carried on the aftermast, the lowest after masthead light shall be at least 4·5 metres vertically higher than the forward masthead light.

(*f*) (i) The masthead light or lights prescribed in Rule 23 (*a*) shall be so placed as to be above and clear of all other lights and obstructions except as described in sub-paragraph (ii).
  (ii) When it is impracticable to carry the all-round lights prescribed by Rule 27 (*b*) (i) or Rule 28 below the masthead lights, they may be carried above the after masthead light(s) or vertically in between the forward masthead light(s) and after masthead light(s), provided that in the latter case the requirement of Section 3 (*c*) of this Annex shall be complied with.

(*g*) The sidelights of a power-driven vessel shall be placed at a height above the hull not greater than three-quarters of that of the forward masthead light. They shall not be so low as to be interfered with by deck lights.

(*h*) The sidelights, if in a combined lantern and carried on a power-driven vessel of less than 20 metres in length, shall be placed not less than 1 metre below the masthead light.

(*i*) When the Rules prescribe two or three lights to be carried in a vertical line, they shall be spaced as follows:
  (i) on a vessel of 20 metres in length or more such lights shall be spaced not less than 2 metres apart, and the lowest of these lights shall, except where a towing light is required, be placed at a height of not less than 4 metres above the hull.
  (ii) on a vessel of less than 20 metres in length such lights shall be spaced not less than 1 metre apart and the lowest of these lights shall, except where a towing light is required, be placed at a height of not less than 2 metres above the hull.
  (iii) when three lights are carried they shall be equally spaced.

(*j*) The lower of the two all-round lights prescribed for a vessel when engaged in fishing shall be at a height above the sidelights not less than twice the distance between the two vertical lights.

(*k*) The forward anchor light prescribed in Rule 30 (*a*) (i), when two are carried, shall not be less than 4·5 metres above the after one. On a vessel of 50 metres or more in length this forward anchor light shall be placed at a height of not less than 6 metres above the hull.

3.   *Horizontal positioning and spacing of lights*

(*a*) When two masthead lights are prescribed for a power-driven vessel, the horizontal distance between them shall not be less than one-half of the length of the vessel but need not be more than 100 metres. The forward light shall be placed not more than one-quarter of the length of the vessel from the stem.

(*b*) On a power-driven vessel of 20 metres or more in length the sidelights shall not be placed in front of the forward masthead lights. They shall be placed at or near the side of the vessel.

(*c*) When the lights prescribed in Rule 27 (*b*) (i) or Rule 28 are placed vertically between the forward masthead light(s) and the after masthead light(s) these all-round lights shall be placed at a horizontal distance of not less than 2 metres from the fore and aft centreline of the vessel in the athwartship direction.

4.   *Details of location of direction-indicating lights for fishing vessels, dredgers and vessels engaged in underwater operations*

(*a*) The light indicating the direction of the outlying gear from a vessel engaged in fishing as prescribed in Rule 26 (*c*) (ii) shall be placed at a horizontal distance of not less than 2 metres and not more than 6 metres away from the two all-round red and white lights. This light shall be placed not higher than the all-round white light prescribed in Rule 26 (*c*) (i) and not lower than the sidelights.

(*b*) The lights and shapes on a vessel engaged in dredging or underwater operations to indicate the obstructed side and/or the side on which it is safe to pass, as prescribed in Rule 27 (*d*) (i) and (ii), shall be placed at the maximum practical horizontal distance, but in no case less than 2 metres, from the lights or shapes prescribed in Rule 27 (*b*) (i) and (ii). In no case shall the upper of these lights or shapes be at a greater height than the lower of the three lights or shapes prescribed in Rule 27 (*b*) (i) amd (ii).

5.   *Screens for sidelights*

The sidelights of vessels of 20 metres or more in length shall be fitted with inboard screens painted matt black, and meeting the requirements of Section 9 of this Annex. On vessels of less than 20 metres in length the sidelights, if necessary to meet the requirements of Section 9 of this Annex, shall be fitted with inboard matt black screens. With a combined lantern, using a single vertical filament and a very narrow division between the green and red sections, external screens need not be fitted.

6.   *Shapes*

(*a*) Shapes shall be black and of the following sizes:
  (i) a ball shall have a diameter of not less than 0·6 metre;
  (ii) a cone shall have a base diameter of not less than 0·6 metre and a height equal to its diameter;
  (iii) a cylinder shall have a diameter of at least 0·6 metre and a height of twice its diameter;
  (iv) a diamond shape shall consist of two cones as defined in (ii) above having a common base.
(*b*) The vertical distance between shapes shall be at least 1·5 metres.

(c) In a vessel of less than 20 metres in length shapes of lesser dimensions but commensurate with the size of the vessel may be used and the distance apart may be correspondingly reduced.

7. *Colour specification of lights*

The chromaticity of all navigation lights shall conform to the following standards, which lie within the boundaries of the area of the diagram specified for each colour by the International Commission on Illumination (CIE).

The boundaries of the area for each colour are given by indicating the corner co-ordinates, which are as follows:

(i) *White*

x	0·525	0·525	0·452	0·310	0·310	0·443
y	0·382	0·440	0·440	0·348	0·283	0·382

(ii) *Green*

x	0·028	0·009	0·300	0·203
y	0·385	0·723	0·511	0·356

(iii) *Red*

x	0·680	0·660	0·735	0·721
y	0·320	0·320	0·265	0·259

(iv) *Yellow*

x	0·612	0·618	0·575	0·575
y	0·382	0·382	0·425	0·406

8. *Intensity of lights*

(a) The minimum luminous intensity of lights shall be calculated by using the formula:

$$I = 3·43 \times 10^6 \times T \times D^2 \times K^{-D}$$

where I   is luminous intensity in candelas under service conditions,

T   is threshold factor 2×10^{-7} lux,

D   is range of visibility (luminous range) of the light in nautical miles,

K   is atmospheric transmissivity.

For prescribed lights the value of K shall be 0·8, corresponding to a meteorological visibility of approximately 13 nautical miles.

(b) A selection of figures derived from the formula is given in the following table:

Range of visibility (luminous range) of light in nautical miles D	Luminous intensity of light in candelas for K = 0·8 I
1	0·9
2	4·3
3	12
4	27
5	52
6	94

Note: The maximum luminous intensity of navigation lights should be limited to avoid undue glare. This shall not be achieved by a variable control of the luminous intensity.

9. *Horizontal sectors*

(a) (i) In the forward direction, sidelights as fitted on the vessel shall show the minimum required intensities. The intensities must decrease to reach practical cut-off between 1 degree and 3 degrees outside the prescribed sectors.

(ii) For sternlights and masthead lights and at 22·5 degrees abaft the beam for sidelights, the minimum required intensities shall be maintained over the arc of the horizon up to 5 degrees within the limits of the sectors prescribed in Rule 21. From 5 degrees within the prescribed sectors the intensity may decrease by 50 per cent up to the prescribed limits; it shall decrease steadily to reach practical cut-off at not more than 5 degrees outside the prescribed sectors.

(*b*) All-round lights shall be so located as not to be obscured by masts, top-masts or structures within angular sectors of more than 6 degrees, except anchor lights prescribed in Rule 30, which need not be placed at an impracticable height above the hull.

10. *Vertical sectors*

(*a*) The vertical sectors of electric lights as fitted, with the exception of light on sailing vessels shall ensure that:
- (i) at least the required minimum intensity is maintained at all angles from 5 degrees above to 5 degrees below the horizontal;
- (ii) at least 60 per cent of the required minimum intensity is maintained from 7·5 degrees above to 7·5 degrees below the horizontal.

(*b*) In the case of sailing vessels the vertical sectors of electric lights as fitted shall ensure that:
- (i) at least the required minimum intensity is maintained at all angles from 5 degrees above to 5 degrees below the horizontal;
- (ii) at least 50 per cent of the required minimum intensity is maintained from 25 degrees above to 25 degrees below the horizontal.

(*c*) In the case of lights other than electric these specifications shall be met as closely as possible.

11. *Intensity of non-electric lights*

Non-electric lights shall so far as practicable comply with the minimum intensities, as specified in the Table given in Section 8 of this Annex.

12. *Manoeuvring light*

Notwithstanding the provisions of paragraph 2 (*f*) of this Annex the manoeuvring light described in Rule 34 (*b*) shall be placed in the same fore and aft vertical plane as the masthead light or lights and, where practicable, at a minimum height of 2 metres vertically above the forward masthead light, provided that it shall be carried not less than 2 metres vertically above or below the after masthead light. On a vessel where only one masthead light is carried, the manoeuvring light, if fitted, shall be carried where it can best be seen, not less than 2 metres vertically apart from the masthead light.

13. *Approval*

The construction of lights and shapes and the installation of lights on board the vessel shall be to the satisfaction of the appropriate authority of the State whose flag the vessel is entitled to fly.

ANNEX II
Additional signals for fishing vessels fishing in close proximity

1. *General*

The lights mentioned herein shall, if exhibited in pursuance of Rule 26 (*d*), be placed where they can best be seen. They shall be at least 0·9 metre apart but at a lower level than lights prescribed in Rule 26 (*b*) (i) and (*c*) (i). The lights shall be visible all round the horizon at a distance of at least 1 mile but at a lesser distance than the lights prescribed by these Rules for fishing vessels.

2. *Signals for trawlers*

(*a*) Vessels when engaged in trawling, whether using demersal or pelagic gear, may exhibit:
- (i) when shooting their nets:
    two white lights in a vertical line;

   (ii) when hauling their nets:
        one white light over one red light in a vertical line;
  (iii) when the net has come fast upon an obstruction:
        two red lights in a vertical line.

  (*b*) Each vessel engaged in pair trawling may exhibit:
   (i) by night, a searchlight directed forward and in the direction of the other
      vessel of the pair;
  (ii) when shooting or hauling their nets or when their nets have come fast
      upon an obstruction, the lights prescribed in 2 (*a*) above.

3. *Signals for purse seiners*

   Vessels engaged in fishing with purse seine gear may exhibit two yellow lights in a vertical line. These lights shall flash alternately every second and with equal light and occultation duration. These lights may be exhibited only when the vessel is hampered by its fishing gear.

ANNEX III

Technical details of sound signal appliances

1. *Whistles*

(*a*) *Frequencies and range of audibility*
   The fundamental frequency of the signal shall lie within the range 70–700 Hz.
   The range of audibility of the signal from a whistle shall be determined by those frequencies, which may include the fundamental and/or one or more higher frequencies, which lie within the range 180–700 Hz ($\pm$ 1 per cent) and which provide the sound pressure levels specified in paragraph 1 (*c*) below.

(*b*) *Limits of fundamental frequencies*
   To ensure a wide variety of whistle characteristics, the fundamental frequency of a whistle shall be between the following limits:
   (i) 70–200 Hz, for a vessel 200 metres or more in length
  (ii) 130–350 Hz, for a vessel 75 metres but less than 200 metres in length;
 (iii) 250–700 Hz, for a vessel less than 75 metres in length.

(*c*) *Sound signal intensity and range of audibility*
   A whistle fitted in a vessel shall provide, in the direction of maximum intensity of the whistle and at a distance of 1 metre from it, a sound pressure level in at least one 1/3rd-octave band within the range of frequencies 180–700 Hz ($\pm$ 1 per cent) of not less than the appropriate figure given in the table below.

Length of vessel in metres	1/3rd-octave band level at 1 metre in dB referred to 2×10^{-5} N/m^2	Audibility range in nautical miles
200 or more	143	2
75 but less than 200 ..	138	1·5
20 but less than 75 ..	130	1
Less than 20	120	0·5

   The range of audibility in the table above is for information and is approximately the range at which a whistle may be heard on its forward axis with 90 per cent probability in conditions of still air on board a vessel having average background noise level at the listening posts (taken to be 68 dB in the octave band centred on 250 Hz and 63 dB in the octave band centred on 500 Hz).

In practice the range at which a whistle may be heard is extremely variable and depends critically on weather conditions; the values given can be regarded as typical but under conditions of strong wind or high ambient noise level at the listening post the range may be much reduced.

(d) Directional properties

The sound pressure level of a directional whistle shall be not more than 4 dB below the prescribed sound pressure level on the axis at any direction in the horizontal plane within $\pm$ 45 degrees of the axis. The sound pressure level at any other direction in the horizontal plane shall be not more than 10 dB below the prescribed sound pressure level on the axis, so that the range in any direction will be at least half the range on the forward axis. The sound pressure level shall be measured in that 1/3rd-octave band which determines the audibility range.

(e) Positioning of whistles

When a directional whistle is to be used as the only whistle on a vessel, it shall be installed with its maximum intensity directed straight ahead.

A whistle shall be placed as high as practicable on a vessel, in order to reduce interception of the emitted sound by obstructions and also to minimize hearing damage risk to personnel. The sound pressure level of the vessel's own signal at listening posts shall not exceed 110 dB (A) and so far as practicable should not exceed 100 dB (A).

(f) Fitting of more than one whistle

If whistles are fitted at a distance apart of more than 100 metres, it shall be so arranged that they are not sounded simultaneously.

(g) Combined whistle systems

If due to the presence of obstructions the sound field of a single whistle or of one of the whistles referred to in paragraph 1 (f) above is likely to have a zone of greatly reduced signal level, it is recommended that a combined whistle system be fitted so as to overcome this reduction. For the purposes of the Rules a combined whistle system is to be regarded as a single whistle. The whistles of a combined system shall be located at a distance apart of not more than 100 metres and arranged to be sounded simultaneously. The frequency of any one whistle shall differ from those of the others by at least 10 Hz.

2.　Bell or gong

(a) Intensity of signal

A bell or gong, or other device having similar sound characteristics shall produce a sound pressure level of not less than 110 dB at a distance of 1 metre from it.

(b) Construction

Bells and gongs shall be made of corrosion-resistant material and designed to give a clear tone. The diameter of the mouth of the bell shall be not less than 300 mm for vessels of 20 metres or more in length, and shall be not less than 200 mm for vessels of 12 metres or more but of less than 20 metres in length. Where practicable, a power-driven bell striker is recommended to ensure constant force but manual operation shall be possible. The mass of the striker shall be not less than 3 per cent of the mass of the bell.

3.　Approval

The construction of sound signal appliances, their performance and their installation on board the vessel shall be to the satisfaction of the appropriate authority of the State whose flag the vessel is entitled to fly.

ANNEX IV

Distress signals

1. The following signals, used or exhibited either together or separately, indicate distress and need of assistance:

(*a*) a gun or other explosive signal fired at intervals of about a minute;

(*b*) a continuous sounding with any fog-signalling apparatus;

(*c*) rockets or shells, throwing red stars fired one at a time at short intervals;

(*d*) a signal made by radiotelegraphy or by any other signalling method, consisting of the group . . . – – – . . . (SOS) in the Morse Code;

(*e*) a signal sent by radiotelephony consisting of the spoken word "Mayday";

(*f*) the International Code Signal of distress indicated by N.C.;

(*g*) a signal consisting of a square flag having above or below it a ball or anything resembling a ball;

(*h*) flames on the vessel (as from a burning tar barrel, oil barrel, etc.);

(*i*) a rocket parachute flare or a hand flare showing a red light;

(*j*) a smoke signal giving off orange-coloured smoke;

(*k*) slowly and repeatedly raising and lowering arms outstretched to each side;

(*l*) the radiotelegraph alarm signal;

(*m*) the radiotelephone alarm signal;

(*n*) signals transmitted by emergency position-indicating radio beacons.

2. The use of exhibition of any of the foregoing signals except for the purpose of indicating distress and need of assistance and the use of other signals which may be confused with any of the above signals is prohibited.

3. Attention is drawn to the relevant sections of the International Code of Signals, the Merchant Ship Search and Rescue Manual and the following signals:

(*a*) a piece of orange-coloured canvas with either a black square and circle or other appropriate symbol (for identification from the air);

(*b*) a dye marker.

CONVERSION FACTORS

The table below shows the main "Imperial" units and the corresponding "SI" units together with a factor with which to multiply the numerical value of the "Imperial" units to give the equivalent numerical value in "SI" units.

"Imperial" Unit	*"SI" Unit*		*Factor*
Nautical Mile	Kilometre	(km)	1·8532 km
Fathom	Metre	(m)	1·82 m
Foot	Metre	(m)	0·305 m
Foot (Draught)	Decimetre	(dm)	3·05 dm
Inch	Millimetre	(mm)	25·4 mm
Square Foot	Square Metre	(m²)	0·093 m²
Square Inch	Square mm	(mm²)	645·2 mm²
Cubic Foot	Cubic Metre	(m³)	0·0283 m³
Gallon	Litre	(l)	4·546 l
Pint	Litre	(l)	0·568 l
Ton	Megagramme or "tonne"	(Mg)	1·016 tonnes
Pound (lb)	Kilogramme	(kg)	0·454 kg
Hundredweight	Kilogramme	(kg)	50·80 kg
Measurement Ton of 40 cubic ft.	Cubic Metre	(m³)	1·133 m³

Note:—

(i) *Tonnages.*—The volume tonnage of vessels (i.e. the Gross and Nett Tonnage) is universally measured in tons of 100 c.ft. and this will continue to be the practice pending international agreement to any change.

Deadweight being a measurement of mass will be in "tonnes" as given above.

The measurement ton of 40 c.ft. as a freight measure could be replaced by the cubic metre and the conversion factor is included above.

(ii) Temperatures will be recorded in degrees Centigrade (°C) though an attempt is being made to call them degrees Celsius (°C).

The conversion from Fahrenheit is given by
$$°C = \tfrac{5}{9}\,(°F - 32)$$

INDEX

INDEX